CIN1277    RC

W9-DJG-186

note stinking

# Programmable Controllers

## Second Edition

# Programmable Controllers

## Thomas A. Hughes

### Second Edition

**Resources for Measurement and Control Series**

**Instrument Society of America**

Copyright © 1997 by   Instrument Society of America
                      67 Alexander Drive
                      P.O. Box 12277
                      Research Triangle Park, NC 27709

Printed in the United States of America.
10 9 8 7 6 5 4 3 2

ISBN 1-55617-610-4

Library of Congress Cataloging-in-Publication Data

Hughes, Thomas A.
   Programmable controllers / Thomas A. Hughes.—2nd ed.
      p.   cm.—(Resources for measurement and control series)
   Includes index.
   ISBN 1-55617-610-4
   1. Programmable controllers.   I. Title.   II. Series
TJ223.P76H84    1997                                    97-2872
629.8'9—dc21                                            CIP

## ISA Resources for Measurement and Control Series (RMC)

- *Measurement and Control Basics,* 2nd Edition (1995)
- *Industrial Level, Pressure, and Density Measurement* (1995)
- *Industrial Flow Measurement* (1990)
- *Programmable Controllers,* 2nd Edition (1997)
- *Control Systems Documentation: Applying Symbols and Identification* (1993)
- *Industrial Data Communications: Fundamentals and Applications,* 2nd Edition (1997)
- *Real-Time Control Networks* (1993)
- *Automation Systems for Control and Data Acquisition* (1992)
- *Evaluation Control Systems Reliability: Techniques and Applications* (1992)
- *Batch Control Systems: Design, Application, and Implementation* (1990)

## THIS BOOK IS DEDICATED TO

my wife Ellen, my daughter Audrey, and the rest of my family,
for their love and encouragement over the years.

# CONTENTS

# PREFACE

Since 1989, this book has been used both as a textbook for Programmable Controller courses and for self-study by thousands of professionals. This applications-based book provides a clear and concise presentation of the fundamental principles of programmable controllers for process and equipment control. This second edition covers all phases of programmable controller applications from design and programming to installation, maintenance, and start-up. This edition also discusses recent advances in programmable controller applications methods such as Graphical Interface Unit (GUI) software, PLC programming standards, and SoftPLCs.

The text provides a complete and comprehensive presentation on the design, programming, documentation, maintenance, troubleshooting, and start-up of several important control applications. The material also includes chapters on numbering systems and binary codes, digital and ladder logic, electrical and electronic principles, input and output systems, memory and storage devices, basic and advanced PLC programming languages, and data communication systems. The chapter on PLC applications has been improved and expanded with additional control application programs. The final chapter covers the installation, maintenance and troubleshooting of programmable controllers in greater detail.

All the chapters have been supplemented with new or improved example problems and exercises. Most of the illustrations in the book have been revised and improved.

I would like to express my appreciation my wife Ellen for the long hours spent reviewing both the first and second edition. I would also like to thank Mr. Thomas Fisher and Mr. Thomas McAvinew for making numerous constructive comments that improved the overall presentation of this second edition.

# ABOUT THE AUTHOR

Thomas A. Hughes, a Senior Member of ISA, has 25 years' experience in the design and application of instrumentation and control systems, including 15 years in the management of control system projects for the process industries. He is the author of two books: *Measurement and Control Basics, 2nd Edition,* (1995) and *Programmable Controllers,* (1997), both published by ISA.

Dr. Hughes received a B.S. in engineering physics from the University of Colorado, and M.S. in control system engineering from Colorado State University, and a Ph.D. in engineering management from California Coast University. He holds professional engineering licenses in the states of Colorado and Alaska, and has held engineering and management positions with Dow Chemical, Stearns-Roger Engineering, Rockwell International, EG&G Rocky Plats, Topro Systems Integration, and Nova Systems. Dr. Hughes has taught numerous courses in electronics, mathematics, and instrumentation and control systems at the college level and in industry. He is currently an Instrumentation Specialist with the International Atomic Energy Agency in Vienna, Austria.

# 1

# Introduction to Programmable Controllers

## Introduction

Programmable controllers were originally designed to replace relay-based control systems and solid-state, hard-wired logic control panels. However, the modern programmable controllers system is far more complex and powerful.

A programmable controller examines the status of inputs and outputs and, in response, controls some process or machine through outputs. These combinations of input and output data are referred to as control logic. Several logic combinations are usually required to carry out a control plan or program. This control plan is stored in memory using a programming device to input the program into the system. The control plan in memory is periodically scanned by the processor, usually a microprocessor, in a predetermined sequential order. The period required to examine the inputs, perform the control logic, and execute the outputs is called the "scan time."

A simplified block diagram of a programmable controller is shown in Figure 1-1, where instruments, such as limit switches and panel-

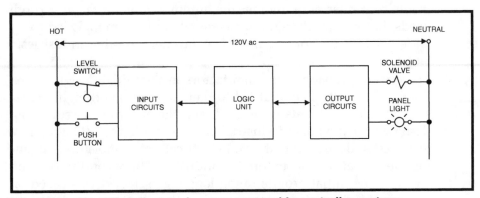

Figure 1-1. Simplified diagram for a programmable controller system.

mounted push buttons, are wired to input circuits, and the output circuits in turn drive devices such as electric solenoid valves and indicator lights. The input and output circuits are controlled by the logic unit.

This figure shows a typical configuration of the early programmable controller applications, which were intended to replace relay or hard-wired logic control systems. The input circuits are used to convert the various field voltages and currents to the low voltage signals [normally 0 to 5 volts direct current (dc)] used by the logic unit. The output circuits convert the logic signals to a form that will drive the field devices. For example, in Figure 1-1, 120 volt ac power is used to power the field devices, so the input and output circuits are used to convert the 0 to 5 volt logic signals to 120 Vac field signals that interface to field devices.

## Brief History of Programmable Controllers

In 1968, a major automobile manufacturer wrote a design specification for the first programmable controller. The primary goal was to eliminate the high cost associated with the frequent replacement of inflexible relay-based control system. The specification also called for a solid-state industrial computer that could be easily programmed by maintenance technicians and plant engineers. It was hoped that the programmable controller would reduce production downtime and provide expandability for future production improvements and changes. In response to this design specification, several manufacturers developed computer-based control devices called programmable controllers.

The first programmable controller was installed in 1969, and it proved to be a vast improvement over relay-based control systems. They were easy to install and program, they used less plant floor space, and were more reliable than a relay-based control system. The initial program-mable controller not only met the automobile manufacturer production needs, but further design improvements in later models led to widespread use of programmable controllers in other industries.

There were probably two main factors in the initial design of pro-grammable controllers that led to their success. First, highly reliable solid-state components were used, and the electronic circuits were designed for the harsh industrial environment. The I/O circuits were designed and built to withstand electrical noise, moisture, oil, and the high temperature encountered in industry. The second important factor was that the initial programming language selected was based on standard electrical ladder logic design. Some earlier computer system applications had failed because plant technicians and engineers were not easily trained in standard computer software. However, most were already trained in relay ladder logic design, so that programming in a language based on the familiar relay ladder diagrams was learned quickly.

When microprocessors were introduced in 1974 and 1975, the basic capabilities of programmable controllers were greatly expanded and improved. They were able to perform sophisticated math and data manipulation function, which greatly increased the use of programmable controller in more complex control applications.

In the late 1970s, improved communication components and circuits made it possible to place programmable controllers thousands of feet from the equipment they controlled, and several programmable controllers can now exchange data to more effectively control processes and machines. Also, microprocessor-based input and output modules allowed programmable controller systems to evolve into the analog control world.

Programmable controllers are found in thousands of industrial applications. They are used to control chemical, petrochemical, food, pharmaceutical, waste water treatment, water treatment, nuclear, natural gas, and mining processes. They are found in material transfer and storage systems that transport and store both the raw materials and the finished products. They are used with robots to perform hazardous industrial operations to allow for safer operations. Programmable controllers are used in conjunction with other computers to perform process and machine data collection and reporting functions, including statistical process control, quality assurance, and online diagnostics. They are utilized in energy management systems to reduce costs and improve environmental control of industrial facilities and office buildings.

The introduction of the personal computer (PC) in the early 1980s greatly increased the power and utility of the programmable controller system in process and machine control. The low cost of personal computers led to their extensive use as programming devices and operator interface control stations. The development of low cost graphical control software packages on personal computers has led to the extensive use of Graphical User Interfaces (GUIs) in programmable controller applications.

Because of the wide use of the personal computer both in control and business applications, the abbreviation PC is generally reserved for personal computers and the abbreviation, PLCs is used for programmable controllers or programmable logic controllers. The abbreviation PLCs will be used in this book to represent programmable controllers.

# Basic Components of PLC Systems

Regardless of size, cost or complexity, all programmable controllers share the same basic components and functional characteristic. A programmable controller will always consist of a processor, an input/output system, a memory unit, a programming language and device, and a power supply. A block diagram of a typical PLC system is shown in Figure 1-2.

## The Processor

The processor consists of one or more standard or custom microprocessors and other integrated circuits that perform the logic and memory functions of the PLC system. The CPU reads the inputs, executes logic as determined by the application program, performs calculations, and controls the outputs accordingly.

There are two areas of memory in the processor that the PLC user can access: program files and data files. Program files store the control application program, subroutine files, and the error file. Data files store data associated with the control program, such as input/output status bits, counter and timer preset and accumulated values, and other stored constants or variables. Together, these two general memory areas are

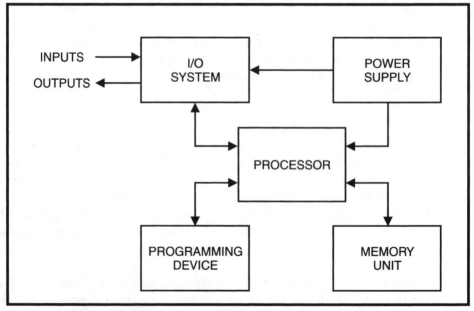

Figure 1-2. Block diagram of a typical PLC system.

called user or application memory. The processor also has an executive or system memory that directs and performs operational activities such as executing the control program and coordinating input scans and output updates. This system memory, which is programmed by the PLC manufacturer, cannot be accessed or changed by the user.

The processor controls the operating cycle or processor scan. This operating cycle consists of a series of operations performed sequentially and repeatedly. A typical PLC processor operating cycle is shown in Figure 1-3.

During the "input scan" the PLC examines the external input devices for a signal present or absent; i.e., an On or Off state. The status of these inputs is temporarily stored in an input image table or memory file. During the "program scan," the processor scans the instructions in the control program, uses the input status from the input image file, and determines if an output will or will not be energized. The resulting status of the outputs is written to the output image table or memory file. Based on the data in the output image table, the PLC energizes or deenergizes its associated output circuits, controlling external devices. This operating cycle typically takes 1 to 25 milliseconds (thousandths of a second).

Since the processor in most programmable controller systems is microprocessor based, we will briefly discuss the basic operation of microprocessors. A microprocessor can be defined as a collection of digital circuits connected together to form an information processing unit. The five essential elements of a microprocessor are shown in Figure 1-4. The five elements are a system clock that synchronizes the actions of the other system components; program memory that stores the program the system will execute; a data memory to store the information that is being manipulated; input/output (I/O) ports for connecting to exterior devices; and a central processing unit (CPU),

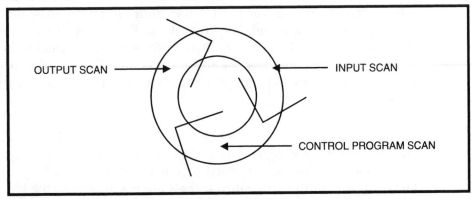

**Figure 1-3. PLC Processor Operating Cycle.**

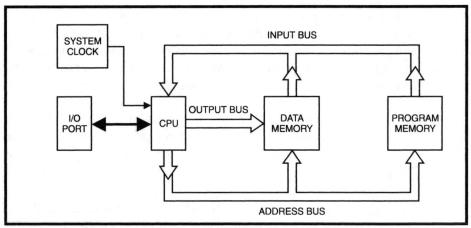

**Figure 1-4. Microprocessor block diagram.**

which operates on the data in the sequence determined by the program. For example, if the system is required to find the sum of a set of numbers, then the numbers will be stored in data memory, the summing program will be stored in the program memory, and the actual calculations will be performed by the CPU.

The CPU is the focal point of the microprocessor, but it must have facilities for storing the control program, and there must be some sort of data storage. The program is stored in memory as a set of binary numbers (0 or 1), which are "coded" to represent the various steps the microprocessor will execute. The microprocessor contains digital logic circuits that decode the program instructions and implement the prescribed program steps.

An important feature of a microprocessor is the manner in which three main system components (CPU, program memory, and data memory) transfer data with each other using buses. A bus is defined as a set of physical connections along which parallel binary information can be transmitted. Only one word of parallel information can exist on a bus at any given time interval. For example, in Figure 1-4, it would not be permissible for the CPU to simultaneously send an address to both the program and data memory via the address bus. The duration of the time interval of any data transfer is dependent of the width of the clock pulse.

The CPU controls the various buses by means of special signals sent out on control lines to the data and program memories. Sometimes these control signals are grouped together under the heading "control bus," but for clarity the control signals are not shown in the block diagram in Figure 1-4.

In executing a program, the microprocessor uses the data buses and control signals to exchange information in a well-defined manner. The complete sequence of information exchange that carries out one program step is known as an "instruction cycle." An example of instruction cycle might consist of the following steps:

1. Fetch the next instruction (i.e., program step) from program memory and place it in the instruction register in the CPU.
2. Increment the program counter by one.
3. Execute the instruction.

The CPU processes program in thousands to millions of steps, each step consists of 8-, 16-, or 32-bit instructions depending on the microprocessor word size. It is important to realize that all microprocessors are sequential machines that operate on 8-, 16-, or 32-bit words, so when we talk about a program scan we may be discussing millions of tiny steps. A word may represent an address or an instruction or a piece of data at a given address. Programs are executed quickly because the clock speeds can be as high as 200 million cycles per second (200 MHz).

Microprocessors are classified as to how powerful they are by their word size and their clock speed. Intel, the manufacturer that invented the microprocessor, continues to develop ever more powerful processors. As shown in Table 1-1, beginning with the model 8085, word size and average clock speed have continued to increase. Today, a Pentium microprocessor manipulates instructions and data in 32-bit words at clock speeds as high or higher than 200 MHz.

Although some high-end, large PLCs use the 80386 or 80486 microprocessors, most operate with less powerful microprocessors. Some small programmable controllers use 8-bit microprocessors running at 4.77 MHz. However, the average small programmable controller is a 16-bit machine operating at 12.5 MHz.

| Microprocessor Type | Word Size | Typical Clock Speed |
|---|---|---|
| 8085 | 8-bit | 1 MHz |
| 8086 | 16-bit | 4.77 MHz |
| 80186 | 16-bit | 8 MHz |
| 80286 | 16-bit | 12.5 MHz |
| 80386 | 32-bit | 33 MHz |
| 80486 | 32-bit | 50 MHz |
| Pentium | 32-bit | 200 MHz and up |

Table 1-1. Intel Microprocessors: Word Size and Typical Clock Speed.

The microprocessor in the programmable controller uses its I/O port to communicate with the outside world. However, since the microprocessor uses low voltage direct current (DC) signals only, there must be I/O circuits or modules to convert the microprocessor low level logic signals to the various signals and voltage levels encountered in the process control environment.

## I/O System

The I/O system provides the physical connection between the process equipment and the microprocessor. This system uses various input circuits or modules to sense and measure physical quantities of the process, such as motion, level, temperature, pressure, flow, and position. Based on the status sensed or values measured, the processor controls various output modules to drive field devices such as valves, motors, pumps, and alarms to exercise control over a machine or a process.

### Input Types

The inputs from field instruments or sensors supply the data and information the processor needs to make logical decisions to control a given process or machine. These input signals from varied devices such as push buttons, hand switches, thermocouple, strain gages, etc., are connected to input modules to filter and condition the signal for use by the processor.

### Output Types

The outputs from the programmable logic controller energize or deenergize control devices to regulate processes or machines. These output signals are control voltages from the output circuits, and they are generally not high power signals. For example, an output module sends a control signal that energizes the coil in a motor starter. The energized coil closes the power contacts of the starter. These contacts then close to start the motor. The output modules are usually not directly connected to the power circuit but rather to devices such as motor starter and heater contactors that apply high power (greater than 10 amps) signals to the final control devices.

### I/O Structure

PLCs are classified as micro, small, medium, and large mainly based on the I/O count. Micro PLCs generally have an I/O count of 32 or less, small PLC have less than 256 I/O points, medium-sized PLCs have an I/O count less than 1024, and large PLCs have an I/O count greater than 1024. Micro PLCs are self-contained units with the processor, power supply, and I/O all in one package. Because they are self-contained, micro PLCs are also called packaged controllers. A modular PLC is one that has separate components or modules that interconnect.

The advantage of a packaged controller is that the unit is smaller, costs less, and it is easy to install. A typical wiring diagram for a Micro PLC is shown in Figure 1-5. An Allen-Bradley Micro 1000 with nine inputs and five outputs is shown. The unit is powered with 120 volts of alternating current (ac) with an internal power supply to operate the internal I/O circuits and the built-in microprocessor, and to generate 24 volts dc for the field input switches and contacts.

## Modular Housing

In small, medium and large PLC systems the various I/O modules are normally mounted in a "universal" modular housing. The term universal in this context means that any module can be inserted into any I/O slot. Modular I/O housings are designed so that the I/O modules can be removed without turning off the ac power or removing the field wiring. Most I/O modules are mounted on printed circuit boards that can be inserted into an I/O housing or a card rack.

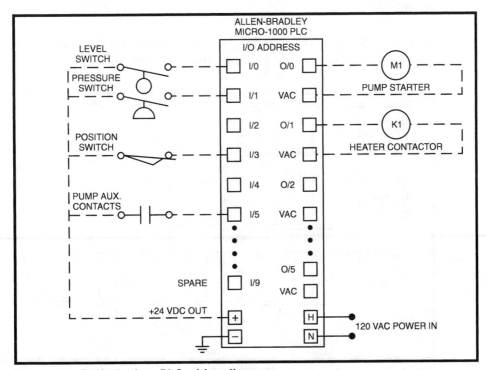

**Figure 1-5. Typical micro-PLC wiring diagram.**

Figure 1-6 shows some typical configurations for I/O modular housings
The backplane of the housings into which the modules are plugged
have a printed circuit card that contains the parallel communications
bus to the processor and the dc voltages to operate the digital and
analog circuits in the I/O modules.

These I/O housings can be mounted in a control panel or on a subpanel in an enclosure. The housings are designed to protect the I/O
module circuits from dirt, dust, electrical noise, and mechanical
vibration.

The backplane of the I/O chassis has sockets for each module. These
sockets provide the power and data communications connection to the
processor for each module. The modules are designed for a wide range
of manufacturing and process applications and include the following
I/O types: discrete, analog, special function, and intelligent.

### Discrete Inputs/Outputs

Discrete is the most common class of input/output in a programmable
controller system. This type of interface module connects field devices
that have two discrete states, such as on/off or open/closed, to the
processor. Each discrete I/O module is designed to be activated by
some field-supplied voltage signal, such as 220V ac, 120 V ac, and
so on.

In a discrete input (DI) module, if an input switch is closed, a electronic
circuit in the input module senses the supplied voltage and converts it

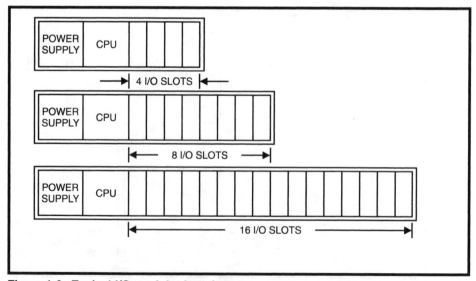

**Figure 1-6. Typical I/O modular housings.**

to a logic-level signal acceptable to the processor to indicate the status of that device. A logic 1 indicates ON or CLOSED, and a logic 0 indicates OFF or OPENED for a field input device or switch.

A typical discrete input module is shown in Figure 1-7. Most input modules will have a light-emitting diode (LED) to indicate the status of each input.

In a discrete output (DO) module, the output interface circuit switches the supplied control voltage that will energize or deenergize the field device. If an output is turned ON through the control program, the supplied control voltage is switched by the interface circuit to activate the referenced (addressed) output device.

Figure 1-8 shows a typical discrete output module wiring diagram. It can be thought of as a simple switch through which power can be provided to control the output device. During normal operation, the processor sends the output state determined by the logic program to the output module. The module then switches the power to the field device.

A fuse is normally provided in the output circuit of the module to prevent excessive current from damaging the wiring to the field device. If the fuse is not provided in the output module, it should be provided in the system design.

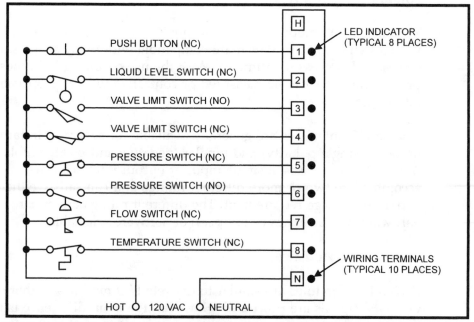

**Figure 1-7. Typical discrete input module wiring diagram.**

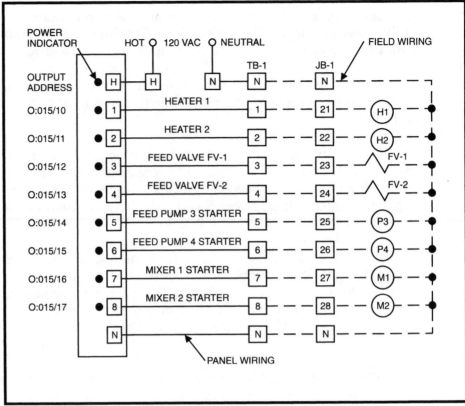

**Figure 1-8. Typical discrete output module wiring diagram.**

## Analog I/O Modules

The analog I/O modules allow for monitoring and controlling of analog voltages and currents, which are compatible with many sensors, motor drives, and process instruments. With the use of analog I/O, most process variables can be measured or controlled with appropriate interfacing.

Analog I/O interfaces are generally available for several standard unipolar (single polarity) and bipolar (negative and positive polarity) ratings. In most cases, a single input or output interface can accommodate two or more different ratings and can satisfy either a current or voltage requirement. The different ratings are either hardware (i.e., switches or jumpers) or software selectable.

## Digital I/O Modules

Digital I/O modules are similar to discrete I/O modules in that discrete ON/OFF signals are processed. However, the main difference is that discrete I/O interfaces require only a single bit to read an input or control an output. On the other hand, digital I/O modules process a group of discrete bits in parallel or serial form.

Typical devices that interface with digital input modules are binary encoders, bar code readers, and thumbwheel switches. Some instruments driven by digital output modules include LED displays, intelligent panels, and BCD displays.

### Special Purpose Modules

The discrete and analog and I/O modules will normally cover about 80% of the input and output signals encountered in programmable controller applications. However, to process certain types of signals or data efficiently, the programmable controller system will require special purpose modules. These special interfaces include those that condition input signals, such as thermocouple modules, pulse counters, or other signals that cannot be interfaced using standard I/O modules. Special purpose I/O modules may also use an on-board microprocessor to add intelligence to the interface. These intelligent modules can perform complete processing functions independent of the CPU and the control program scan.

Another important class of special purpose I/O modules are communication modules that communicate with distributed control systems (DCSs), other PLCs networks, plant computers, or other intelligent devices.

## Memory

Memory is used to store the control program for the PLC system; it is usually located in the same housing as the CPU. The information stored in memory determines how the input and output data will be processed.

Memory stores individual pieces of data called bits. A bit has two states: 1 or 0. Memory units are mounted on circuit boards and are usually specified in thousands or "K" increments where 1K is 1024 words (i.e., $2^{10} = 1024$) of storage space. Programmable controller memory capacity may vary from less than one thousand words to over 64,000 words (64K words) depending on the programmable controller manufacturer. The complexity of the control plan will determine the amount of memory required.

Although there are several different types of computer memory, they can always be classified as *volatile* or *nonvolatile*. Volatile memory will lose its programmed contents if all operating power is lost or removed.

Volatile memory is easily altered and quite suitable for most programming applications when supported by battery backup and/or a recorded copy of the program. Nonvolatile memory will retain its data and program even if there is a complete loss of operating power. It does not require a backup system.

The most common form of volatile memory is Random Access Memory, or RAM. RAM is relatively fast and provides an easy means to create and store application programs. If normal power is disrupted, PLCs with RAM memory use battery or capacitor backups to prevent program loss.

The Electrically Erasable Programmable Read Only (EEPROM) Memory is a nonvolatile memory that is programmed through application software, which runs on a personal computer or through a micro PLC handheld programmer.

## Programming Languages

The programming language allows the user to communicate with the programmable controller via a programming device. Programmable controller manufacturers use several different programming languages, but they all convey to the system, by means of instructions, a basic control plan.

A control plan or program is defined as a set of instructions that are arranged in a logical sequence to control the actions of a process or machine. For example, the program might direct the programmable controller to turn a motor starter on when a push button is depressed and at the same time direct the programmable controller to turn on a control panel-mounted RUN light when the motor starter auxiliary contacts are closed.

A program is written by combining instructions in a certain order. Rules govern the manner in which instructions are combined and the actual form of the instructions. These rules and instructions combine to form a language.

The four most common types of languages encountered in programmable controllers in the United States of America are as follows:

1. Ladder logic.
2. Structured text (such as Boolean logic).
3. Ladder Logic with advanced function blocks.
4. Sequential function chart.

## Ladder Logic

Ladder Logic is the most common PLC language. The reason for this is relatively simple. The original programmable controllers were designed to replace electrical relay-based control systems. These systems were designed by technicians and engineers using a symbolic language called ladder diagrams. The ladder diagram consists of a series of symbols interconnected by lines to indicate the flow of current through the various devices. The ladder drawing consists of basically two things: first is the power source, which forms the sides of the ladder (rails), and the second is the current that flows through the various logic input devices that form the rungs of the ladder.

In electrical design, the ladder diagram is intended to show only the circuitry necessary for the basic operation of the control system. Another diagram, called the wiring diagram, is used to show the physical connection of control devices. The discrete I/O module diagrams shown earlier are examples of wiring diagrams. A typical electrical ladder diagram is shown in Figure 1-9. In this diagram, a push button (PB1) is used to energize a pump start relay (CR1) if the level in a liquid storage tank is not high. Each device has a special symbol assigned to it to make reading of the diagram easier and faster.

The same control application can be implemented using the programmable controller ladder diagram program as shown in Figure 1-10. The diagrams are read in the same manner from left to right, with the logic input conditions on the left and the logical outputs on the right. In the case of electrical diagrams, there must be electrical continuity to energize the output devices; for programmable controller ladder programs, there must be logic continuity to energize the outputs.

In ladder programs, three basic instructions are used to form the program. The first symbol is similar to the normally open (NO) relay

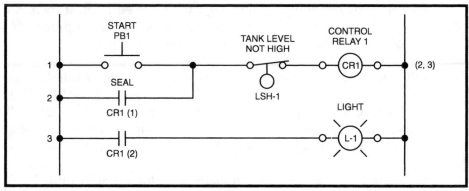

**Figure 1-9. Typical electrical ladder diagram.**

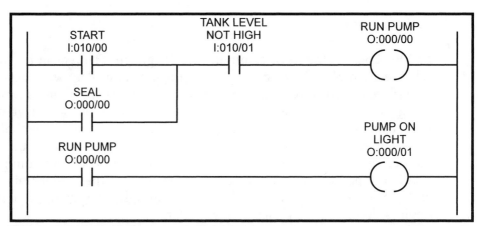

**Figure 1-10. Typical ladder logic PLC program.**

contacts used in electrical ladder diagrams; this instruction uses the same NO symbol in ladder programs. It instructs the processor to examine its assigned bit location in memory. If the bit is ON (logic 1), the instruction is true and there is logic continuity through the instruction on the ladder rung. If the bit is OFF (logic 0), there is no logic continuity through the instruction on the rung. The normally open instruction is also called the examine on instruction by some PLC manufacturers and has the mnemonic symbol of XIC.

The second important instruction is similar to the normally closed (NC) contact from electrical ladder diagrams and it is called the examine off (XIO) instruction. Unlike the examine on instruction, it directs the processor to examine the bit for logical 0 or the OFF condition. If the bit is OFF, the instruction is true and there is logic continuity through the instruction. If the bit is ON, the normally closed instruction is false and there is no logic continuity.

The third instruction is the output coil or energized instruction (OTE). This instruction is similar to relay coil in electrical ladder diagrams and it directs the processor to set a certain location in memory to ON or 1, if there is logic continuity in any logic path preceding it. If there are no complete logic continuity paths in the ladder rung, the processor sets the output coil instruction bit to 0 or OFF.

In Figure 1-10, the letters I or O followed by a 5-digit number above the instructions are the reference addresses for the logic bits. The letter I before the 5-digit number indicates an input bit and the letter "O" before the 5-digit number indicates an output bit. The reference address indicates where in the memory the logic operation will take place.

In Figure 1-10, the examine on instruction for the start push button (PB) directs the processor to see if the reference address I:010/00 is ON. In the same manner, the examine on instruction for the "tank level not high" input instructs the processor to see if the reference address I:010/01 is OFF. If there is logic continuity through both instructions, the output coil or energized (OTE) instruction at address O: 000/00 is turned ON. This same output bit is then used to "seal in" the start push button instruction and also turns on the energized instruction bit O:000/01 to turn on the pump run light.

## Programming Devices

The programming device is used to enter, store, and monitor the programmable controller software. Programming devices can be dedicated portable unit- or personal computer-based system. The personal computer-based systems normally have basic components: keyboard, visual display or CRT, personal computer, printer, and communications interface card and cable as shown in Figure 1-11.

The programming devices are normally connected to the programmable controller system only during programming, startup, or troubleshooting of the control system. Otherwise the programming device is disconnected from the system.

The programming terminals are normally either a handheld programmer or a personal computer based system. The handheld programmers are inexpensive and portable units, normally used to program small programmable controllers. Most of these units resemble portable calculators, but with larger displays and a somewhat different keyboard. The displays are generally LED (light-emitting diode) or dot matrix LCD (liquid crystal display), and the keyboard consists of alphanumeric keys, programming instruction keys, and special function keys. Even though they are mainly used for writing and editing the

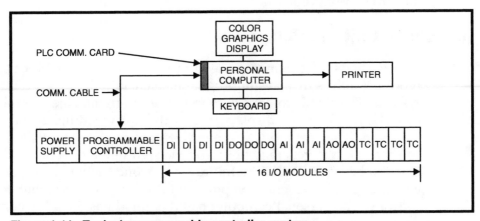

**Figure 1-11. Typical programmable controller system.**

control program, the portable programmers are also used for testing, changing, and monitoring the program.

The standard programming terminal is a personal computer as shown in Figure 1-11 with the programming software loaded on the hard drive. These units can perform program editing and storage. They also have added features such as automatic program printouts and connection to local area networks (LAN). LANs give the programmer or engineer access to any programmable controller in the communications network, so that any device in the network can be monitored and controlled.

## Power Supply

The power supply converts ac line voltages to dc voltages to power the electronic circuits in a programmable controller system. These power supplies rectify, filter, and regulate voltages and currents to supply the correct amounts of voltage and current to the system. The power supply normally converts 120 V ac or 240 V ac line voltage into direct current (dc) voltages such as $+5$ Vdc, $-15$ Vdc, or $+15$ Vdc.

The power supply for a programmable controller system may be integrated with the processor, memory, and I/O modules into a single housing, or it might be a separate unit connected to the system through a cable. As a system expands to include more I/O modules or special function modules, most programmable controllers require an additional or auxiliary power supply to meet the increased power demand. Programmable controller power supplies are usually designed to eliminate electrical noise present on the ac power and signal lines of industrial plants so that this electrical noise does not introduce errors in the control system. They are also designed to operate properly in the higher temperature and humidity environments present in most industrial applications.

## Graphical Unit Interface

Two methods are commonly used to provide operators with color process graphics displays in programmable controller-based control systems. The first is to hard wire from the programmable controller I/O modules to a graphics display panel with hard wired lights and digital indicators. This method is cost effective in a small system that will not be changed. It is generally not recommended in larger control systems that will be expanded in the future. The second method is to use an industrial grade personal computer with process color graphics software. The personal computer-based method has the advantages that the process display screens can be easily modified for process changes, and the computer can perform other functions such as alarm listing, report generation, and programmable controller software.

These features can be explained best by using a personal computer-based Graphical Unit Interface (GUI) system on a typical process control system as shown in Figure 1-12.

The software for vendor-supplied GUI color displays is normally menu driven and relatively easy to use. The process display screens are usually based on the process and instrument drawings for the process being controlled.

For example, if we built a process display based on a dehydration process, the computer-generated display would be as shown in Figure 1-12. An advantage of a PC-based GUI is that process data and alarm messages can be displayed on the screen.

In the GUI graphics of Figure 1-12, Tower 1 is shown in service so that the Process Gas Inlet valve FV-1 and the Process Gas Outlet valve FV-7 are open and the Heating Gas Inlet valve FV-3 and the Heating Gas

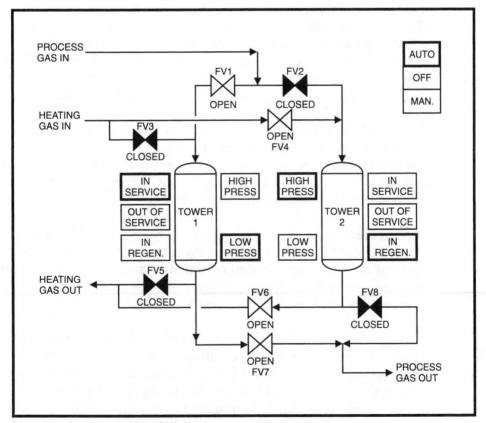

**Figure 1-12. Dehydration GUI display.**

Outlet Valve FV-5 are OFF or closed for Tower 1. To represent the closed condition of the valves, we can fill the valve with a dark color. At the same time, Tower 2 is shown in regeneration with the appropriate valves labeled Open or Closed and the closed valves are filled in with a dark color. We can also display process conditions, such as Tower pressure high or low, on the graphics screen as shown in Figure 1-12.

Another important advance in programmable controller is the use of personal computers to directly replace PLC processors in applications. Generally the standard PLC input and output modules are controlled by an industrial grade personal computer. These systems are generally called SoftLogic or SoftPLC systems.

## SoftLogic or SoftPLC

The original hard logic PLC had the advantage of being designed for the harsh industrial environment; however, with the recent design of industrial grade personal computer this advantage has been lost.

A disadvantage of most PLCs is that they use proprietary software. Proprietary-controlled software suffers from an inherent performance disadvantage. With personal computers, faster and more powerful capabilities are introduced at rapid and regular intervals. Personal computer speeds have historically doubled every 18 months. This is not true of PLCs, which require additional engineering to adapt advanced microprocessor technology to their proprietary architectures.

As a result, it is estimated that hard logic PLC performance lags personal computer industry advances by 18 months to two years, and the gap is widening. With PC processing power leapfrogging itself at this rate, PLCs are having great difficulty keeping up with commodity PC retailing to consumers for under $1,000.

After a decade of exponential growth in capabilities, PCs now come in rugged packages, with ever faster processors, real-time operating systems to handle time-critical operations, larger memory, a multitude of Windows-based software products. All this at ever lower cost, due to competition and high volume manufacturing.

The personal computer is starting to replace the PLC processor in small- and medium-sized automation projects. Some PLC manu-facturers are actually using PC circuit cards inside their PLCs and calling the system a softPLC. The future appears strong for the increased use of personal computers to directly replace proprietary PLCs.

## EXERCISES

1.1 Explain the operation and purpose of each component in a typical programmable controller system.

1.2 List some typical examples of discrete and analog input and output signals found in process industries.

1.3 Explain the difference between volatile and nonvolatile memory.

1.4 Discuss the function and purpose of the various input/output modules used in programmable controller systems.

1.5 List some common applications for personal computers in programmable controller systems.

1.6 Discuss the various types of programming terminals and devices used in programmable controller systems.

1.7 Discuss the basic instructions used in ladder logic programs.

1.8 Discuss the advantages of GUI software in programmable control applications.

## BIBLIOGRAPHY

1. Burton, D. P., and Dexter, A. L., *Microprocessor Systems Handbook*, Analog Devices Inc., 1979.
2. Peatman, J. B., *Microprocessor-Based Design*, McGraw-Hill Inc., 1977.
3. Jones, C. T., and Bryan, L. A., *Programmable Controllers Concepts and Applications*, International Programmable Controllers, Inc., First Edition, 1983.
4. *PLC-2/30 Programmable Controller: Programming and Operations Manual*, Allen-Bradley, Publications 1779-6.8.3, 1984.
5. Gilbert, R. A., and Llewellyn, J. A., *Programmable Controllers — Practices and Concepts*, Industrial Training Corporation, 1985.
6. Plato Computer-Based Training, *Programmable Controller Fundamentals*, Allen-Bradley Company, Inc., 1985.
7. Webb, J. W., and Reis, R. A., *Programmable Logic Controllers — Principles and Applications*, Third Edition, Prentice-Hall, Inc., 1995.
8. *Micro Mentor-Understanding and Applying Micro Programmable Controllers*, Allen-Bradley Company, Inc., 1995.
9. Wisnosky, D. E., *SoftLogic: Overcoming Funnel Vision*, Wizdom Controls, Inc., 1996.

# 2

# Numbering Systems and Binary Codes

## Introduction

The application of programmable controllers to process control requires an understanding of numbering systems and binary codes. Programming, system communications, and input/output (I/O) interfacing in particular require a solid background in numbering systems and binary codes.

## Discrete Signals

The use of digital computers in process control requires that the measurement and control signals be encoded into discrete form. Discrete signals are simply two-state (binary) signals (on/off, start/stop, high voltage/low voltage, etc.).

The simplest approach to encoding analog data into a digital word is provided by the American Standard Code for Information Interchange (ASCII). This method uses a pattern of seven bits (ones and zeros) to represent letters and numbers. Sometimes an extra bit (parity bit) is used to check that the correct pattern has been transmitted. Some examples of coding data using the ASCII standard are given below:

$$
\begin{aligned}
0 &= 011\ 0000 \\
1 &= 011\ 0001 \\
2 &= 011\ 0010 \\
&\quad\ \cdot \\
&\quad\ \cdot \\
&\quad\ \cdot \\
A &= 100\ 0001 \\
B &= 100\ 0010 \\
C &= 100\ 0011 \\
D &= 100\ 0100 \\
&\ \text{etc.}
\end{aligned}
$$

The reader should refer to an ASCII Code Chart (see Table 2-8) in this chapter for a complete listing of conversions between binary strings and the ASCII characters.

Many methods are used for encoding digital numbers in programmable controllers and digital computers. The most common method is the simple binary code that will be discussed later. But first, we will review numbering systems encountered in PLC applications.

# Numbering Systems

The most commonly used numbering system in programmable controllers is the binary system, but the octal and hexadecimal numbering systems are also encountered in these applications. We will start with a brief review of the decimal numbering system, followed by coverage of the other three systems.

## Decimal Numbering System

The decimal numbering system is in common use probably because man started to count with his fingers. However, the decimal system is not easy to implement electronically. A ten-state electronic device would be quite costly and complex. It is much easier and more efficient to use the binary (two-state) numbering system when manipulating numbers using logic circuits.

A decimal number, $N_{10}$ can be written mathematically as:

$$N_{10} = d_n R^n + \cdots + d_2 R^2 + d_1 R^1 + d_0 R^0 \qquad (2\text{-}1)$$

where $R$ is equal to the number of digit symbols used in the system. $R$ is called the radix and is equal to 10 in the decimal system. The subscript 10 on the number $N$ in Equation 2-1 indicates that it is a decimal number. However, it is common practice to omit this subscript in writing out decimal numbers. The decimal digits, $d_n \ldots d_2, d_1, d_0$, can assume the values of 0, 1, 2, 3, 4, 5, 6, 7, 8, or 9 in the decimal numbering system.

For example, the decimal number 1735 can be written as:

$$1735 = 1 \times 10^3 + 7 \times 10^2 + 3 \times 10^1 + 5 \times 10^0$$
$$1735 = 1000 + 700 + 30 + 5$$

When written as 1735, the powers of 10 are implied by positional notation. The value of the decimal number is computed by multiplying each digit by the weight of its position and summing the result. As we will see, this is true for all numbering systems; the decimal equivalent of any number can be calculated by multiplying the digit times its base raised to the power of the digit's position. The general equation is given below:

$$N_b = Z_n R^n + \cdots + Z_2 R^2 + Z_1 R^1 + Z_0 Z^0 \qquad (2\text{-}2)$$

Where $Z$ is the value of the digit, $R$ is the radix, and $b$ is the base of the numbering system.

## Binary Numbering System

The binary numbering system uses the number 2 as the base, and the only allowable digits are 0 or 1. This is the basic numbering system for computers and programmable controllers, which are basically electronic devices that manipulate 0s and 1s to perform math and control functions.

It was easier and more convenient to design digital computers that operate on two entities or numbers rather than the ten numbers used in the decimal world. Furthermore, most physical elements in the process environment have only two states, such as a pump on or off, a valve open or closed, a switch on or off, and so on.

A binary number follows the same format as a decimal one: The value of a digit is determined by where it stands in relation to the other digits in a number. In the decimal system, a 1 by itself is worth 1; placing it to the left of a zero makes the 1 worth 10, and putting it to the left of two zeros makes it worth 100. This simple rule is the foundation of the numbering systems. For example, numbers to be added or subtracted are first arranged so that their place columns line up.

In the decimal system, each position to the left of the decimal point indicates an increasing power of 10. In the binary system, each place to the left signifies an increased power of two, i.e., $2^0$ is one, $2^1$ is two, $2^2$ is four, $2^3$ is eight, and so on. So, finding the decimal equivalent of a binary number is simply a matter of noting which place columns the binary 1s occupy and adding up their values. A binary number also uses standard positional notation and is written as $b_n \ldots b_2, b_1, b_0$. The decimal equivalent of a binary number can be found using the equation:

$$N_2 = Z_n 2^n + \cdots + Z_2 2^2 + Z_1 2^1 + Z_0 2^0 \qquad (2\text{-}3)$$

where the radix $R$ equals 2 in the binary system, and each binary digit (bit) can take on the value of 0 or 1 only.

The decimal equivalent of the binary number 10101 can be found as follows:

$$10101_2 = 1 \times 2^4 + 0 \times 2^2 + 1 \times 2^2 + 0 \times 2^1 + 1 \times 2^0$$

or

$$(1 \times 16) + (0 \times 8) + (1 \times 4) + (0 \times 2) + (1 \times 1) = 21 \text{ (decimal)}$$

---

**EXAMPLE 2-1**

**Problem:** Convert the binary number $101110_2$ to its decimal equivalent.

**Solution:** Using Equation 2-3:

$$N_2 = Z_n 2^n + \cdots + Z_2 2^2 + Z_1 2^1 + Z_0 2^0$$

or

$$1 \times 2^5 + 0 \times 2^4 + 1 \times 2^3 + 1 \times 2^2 + 1 \times 2^1 + 0 \times 2^0$$
$$= (1 \times 32) + (0 \times 16) + (1 \times 8) + (1 \times 4) + (1 \times 2) + (0 \times 1)$$
$$= 32 + 0 + 8 + 4 + 2 + 0$$
$$= 46$$

---

To convert a decimal number to a binary number, we first subtract the largest possible power of two numbers from the decimal number and continue subtracting the next largest power of 2 from the remainder, placing 1s in each column where a subtraction is performed and 0s where there is no subtraction possible. For example, we can convert the decimal number 50 to a binary number as shown in Figure 2-1.

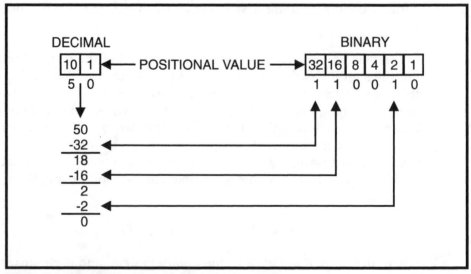

**Figure 2-1. Decimal to binary conversion example.**

## Binary Arithmetic

The rules of addition for binary numbers are illustrated in Table 2-1. As in decimal addition, 0 plus 0 is 0 in binary. Similarly, 0 plus 1 equals 1, as it does in the decimal system. However, the third possibility of binary addition is different from decimal in that 1 plus 1 equals 0, with a carry of 1. In fact, the addition of numbers is performed in accordance with certain rules, regardless of which base (radix) is used to represent the numbers.

When adding two binary numbers, the following algorithm is used:

1. Align the numbers so that their place or positional values correspond.
2. Starting with the least significant digit, add the pair of digits in this column.
3. If the sum is less than or equal to 1, it is expressed as a single binary digit and there is no carry.
4. If the sum of the two digits is greater than 1, it is expressed as a two-digit binary number. The more significant digit, 1, is carried to the next positional column, and the less significant digit is placed under the current column.
5. The digits in the next column, including any carry, are added until all columns have been computed.

---

**EXAMPLE 2-2**

**Problem:** Convert the decimal number 59 to a binary number.

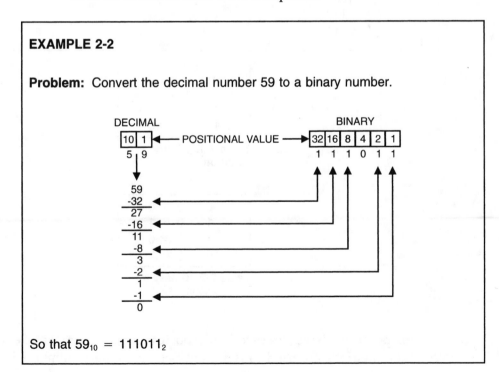

So that $59_{10} = 111011_2$

---

**EXAMPLE 2-3**

**Problem:** Perform binary addition of the following numbers: 1010 and 0110.

**Solution:**

$$
\begin{array}{r}
\text{Carry digits} \quad 111 \\
1010 \\
+ \ 0110 \\
\hline
10000
\end{array}
$$

The rules of binary subtraction are given in Table 2-2.

Note that when we subtract 1 from 0, we must borrow a 1 from the next higher-order bit. The following example problem will help illustrate binary subtraction.

**EXAMPLE 2-4**

**Problem:** Subtract the binary number $110_2$ from the number $1001_2$.

**Solution:**

$$
\begin{array}{r}
1001 \\
- \ 0110 \\
\hline
0011
\end{array}
$$

$$
\begin{array}{rcl}
0 + 0 &=& 0 \\
0 + 1 &=& 1 \\
1 + 0 &=& 1 \\
1 + 1 &=& 10
\end{array}
$$

**Table 2-1. Binary Addition.**

$$
\begin{array}{rcl}
0 - 0 &=& 0 \\
1 - 0 &=& 1 \\
1 - 1 &=& 0 \\
10 - 1 &=& 1
\end{array}
$$

**Table 2-2. Binary Subtraction.**

To this point, we have discussed only positive binary numbers. Several common methods are used to represent negative binary numbers in programmable controller systems. The first is *signed-magnitude binary*. This method places an extra bit (sign bit) in the leftmost position and lets this bit determine whether the number is positive or negative. The number is positive if the sign bit is 0 and negative if the sign bit is 1. For example, in a 16-bit machine, if we have a 12-bit binary number, $00000010101_2 = 21_{10}$, to express the positive and negative values, we would manipulate the leftmost or most significant bit. So, using the signed magnitude method: $000000000010101 = +21$ and $1000000000010101 = -21$.

Another common method used to express negative binary numbers is called *two's complement binary*. To complement a number means to change it to a negative number. For example, the binary number 10101 is equal to decimal 21. To get the negative using the two's complement method, you complement each bit and then added to the least significant bit.

In the case of the binary number $010101 = 21$, its two's complement would be: $101011 = -21$.

## Octal Numbering System

The binary numbering system requires substantially more digits to express a number than the decimal system. For example, $130_{10} = 10000010_2$, so it takes 8 or more binary digits to express a decimal number over 127. It is also difficult for people to read and manipulate large numbers without making errors. To reduce errors in binary number manipulations, some computer manufacturers started using the octal numbering system. This system uses the number 8 as a base with the 8 digits of 0, 1, 2, 3, 4, 5, 6, 7. Like all other number systems, each digit in an octal number has a weighted value according to its position. For example:

$$1301_8 = 1 \times 8^3 + 3 \times 8^2 + 0 \times 8^1 + 1 \times 8^0$$
$$= 1 \times 512 + 3 \times 64 + 0 \times 8 + 1 \times 1$$
$$= 512 + 192 + 0 + 1$$
$$= 705_{10}$$

The octal system is used as a convenient means of writing or manipulating binary numbers in PLC systems. A binary number with a large number of 1s and 0s can be represented with an equivalent octal number with fewer digits. As shown in Table 2-3, one octal digit can be used to express three binary digits so that the number is reduced by a factor of three.

| Binary | Octal |
|--------|-------|
| 000 | 0 |
| 001 | 1 |
| 010 | 2 |
| 011 | 3 |
| 100 | 4 |
| 101 | 5 |
| 110 | 6 |
| 111 | 7 |

**Table 2-3. Binary and Octal Equivalent Numbers.**

For example, the binary number $11010101010_2$ can be converted to an octal number by grouping binary bits in groups of three starting with the least significant bit, as follows:

$$011\ 010\ 101\ 010\ =\ 3252_8$$

---

**EXAMPLE 2-5**

**Problem:** Represent the binary number $101011001101111_2$ in octal.

**Solution:** To convert from a binary to an octal number, we simply divide the binary number into groups of three bits, starting with the least significant bit (LSB). We then use Table 2-3 to convert the 3-bit groups to their octal equivalent. To solve, we place binary number into groups of three: 101 011 001 101 $111_2$. Since $101_2 = 5_8$, $011_2 = 3_8$ , $001_2 = 1_8$, $101_2 = 5_8$, and $111_2 = 7_8$, we obtain $111_2 = 53157_8$.

---

We can convert a decimal number to an octal number by successively dividing the decimal number by the octal base number 8. This is best illustrated in the following example.

---

**EXAMPLE 2-6**

**Problem:** Convert the decimal number $370_{10}$ to an octal number.

**Solution:** Decimal to octal conversion is obtained by successive division by the octal base number 8 as follows:

| Division | Quotient | Remainder |
|----------|----------|-----------|
| 370/8 | 46 | 2 (LSD) |
| 46/6 | 5 | 6 |
| 5/8 | 0 | 5 (MSD) |

So that $370_{10} = 562_8$.

---

## Hexadecimal Numbering System

The hexadecimal numbering system provides an even shorter notation than the octal system and is a commonly used numbering system in PLC applications. The hexadecimal system has a base of 16, and four binary bits are used to represent a single symbol. The sixteen symbols are 0, 1, 2, 3, 4, 5, 6, 7, 8, 9, A, B, C, D, E, and F. The letters A through F are used to represent the binary strings 1010, 1011, 1100, 1101, 1110, and 1111, which correspond to the decimal numbers 10 through 15. The hexadecimal digits and their binary equivalents are given in Table 2-4.

| Hexadecimal | Binary | Hexadecimal | Binary |
|:-----------:|:------:|:-----------:|:------:|
| 0 | 0000 | 9 | 1001 |
| 1 | 0001 | A | 1010 |
| 2 | 0010 | B | 1011 |
| 3 | 0011 | C | 1100 |
| 4 | 0100 | D | 1101 |
| 5 | 0101 | E | 1110 |
| 6 | 0110 | F | 1111 |
| 7 | 0111 | 10 | 10000 |
| 8 | 1000 | 11 | 10001 |

**Table 2-4. Hexadecimal and Binary Equivalent Numbers.**

To convert a binary number to a hexadecimal number, we use Table 2-4. For example, the binary number 0110 1111 1000 is 6F8. Again, the hexadecimal numbers follow the standard positional convention $H_n \dots H_2$, $H_1$, $H_0$ where the positional weights for hexadecimal numbers are powers of sixteen with 1, 16, 256, and 4096 being the first four decimal values.

To convert from hexadecimal to decimal numbers, we use the following equation:

$$N_{16} = H_n 16^n + \dots + H_2 16^2 + H_1 16^1 + H_0 16^0 \qquad (2\text{-}4)$$

where the radix $R$ equals 16 in the hexadecimal system, and each digit can take on the value of 0 through 9 and the letters A, B, C, D, E, and F.

---

**EXAMPLE 2-7**

**Problem:** Convert the hex number 1FA to its decimal equivalent.

**Solution:** $N_{16} = H_n 16^n + \dots + H_2 16^2 + H_1 16^1 + H_0 16^0$

Since $H_2 = 1 = 1_{10}$, $H_1 = F = 15_{10}$, and $H_0 = A = 10_{10}$,

we obtain

$1 \times 16^2 + 15 \times 16^1 + 10 \times 16^0 = 256 + 240 + 10 = 506_{10}$

---

To convert a decimal number to a hexadecimal number, we use the following procedure:

1. Divide the decimal number by 16 and record the quotient and the remainder.
2. Divide the quotient from the division in Step 1 by 16 and record the quotient and the remainder.
3. Repeat Step 2 until the quotient is zero.
4. The hexadecimal equivalents of the remainders generated by the divisions are the digits of the hexadecimal number, where the first remainder is the least significant digit (LSD) and the last remainder is the most significant digit (MSD).

An example will illustrate the conversion from a decimal to hexadecimal number.

---

**EXAMPLE 2-8**

**Problem:** Convert the decimal number $610_{10}$ to a hexadecimal number.

**Solution:**

| Division | Quotient | Remainder |
|----------|----------|-----------|
| 620/16 | 38 | $2_{10} = 2_{16}$ (LSD) |
| 38/16 | 2 | $6_{10} = 6_{16}$ |
| 2/16 | 0 | $2_{10} = 2_{16}$ |

Therefore, $610_{10} = 262_{16}$.

We can check the answer by converting $262_{16}$ back to a decimal number as follows:

$$\begin{aligned} N_{16} &= 2 \times 16^2 + 6 \times 16^1 + 2 \times 16^0 \\ &= 2 \times 256 + 6 \times 16 + 2 \times 1 \\ &= 512 + 96 + 2 = 610_{10} \end{aligned}$$

---

## Hexadecimal Arithmetic

The addition of numbers is performed according to certain rules, regardless of which base (radix) is used to represent the numbers. When adding two hexadecimal numbers, the following algorithm is used:

1. Align the numbers so that their place positional values correspond.

2. Starting with the least significant digit, add the pair of digits in this column.
3. If the sum is less than 16 decimals, it is expressed as a single hexadecimal digit and there is no carry.
4. If the sum of the two digits is greater than or equal to decimal sixteen, it is expressed as a two-digit hexadecimal number. The more significant digit, 1, is carried to the next positional column, and the less significant digit is placed under the current column.
5. The digits in the next column, including any carry, are added until all columns have been computed.

---

**EXAMPLE 2-9**

**Problem:** Add the hexadecimal numbers $1736_{16}$ and $1AC_{16}$ together.

**Solution:** First, align the numbers so that the positional values correspond.

$$\begin{array}{r} 1736 \\ + \ 1AC \\ \hline \end{array}$$

Starting with the right most ($16^0$) positional value, add the pair of digits in this column.

$$\begin{array}{r} 1 \quad \text{Carry Bit}\\ 1736 \\ + \ 1AC \\ \hline 2 \end{array}$$

Since the sum of C + 6 is more than $15_{10}$, there is a carry that must be add to the column just to the left of the right most column.

We now continue with the next column. Here we have 1 (the carry) + 3 + A = E, so there is no carry to the last column.

$$\begin{array}{r} 1736 \\ + \ 1AC \\ \hline E2 \end{array}$$

Finally, we continue in the same manner, adding the last two columns.

$$\begin{array}{r} 1736 \\ + \ 1AC \\ \hline 18E2 \end{array}$$

The subtraction of numbers is performed using the same process for all numbering systems. It is basically a reversal of the addition process. The most convenient method for subtracting a hexadecimal number is to convert each digit above 9 to a decimal, use standard decimal subtraction, and then convert the decimal result back to hexadecimal. For example, confronted with the problem $F - A$, convert to $15 - 10$ to arrive at the solution, as follows:

$$\begin{array}{ll} & F \quad (15_{10}) \\ - & A \quad (10_{10}) \\ \hline & 5_{16} \quad (5_{10}) \end{array}$$

Borrowing is performed in the same manner as in the decimal numbering system. But we must remember that a borrow of 1 in the hexadecimal system is equivalent to 16 in the decimal system.

---

**EXAMPLE 2-10**

**Problem:** Subtract $29_{16}$ from $68_{16}$.

**Solution:** To solve the problem of $68_{16} - 29_{16}$, we first note that 8 is less than 9, so we must borrow a 1 from the $16^1$ column. Solving the problem in decimal, the value borrowed is $16_{10}$, so that $16_{10} + 8_{10} - 9_{10} = 24_{10} - 9_{10} = 15_{10} = F_{16}$ is the result for the right-hand column.

$$\begin{array}{r} 68 \\ - 29 \\ \hline F \end{array}$$

Next, we have $5 - 2 = 3$ in the left column, where the 6 has been reduced to 5, since we needed to borrowed from this column to perform the subtraction in the right-hand column. The result is

$$\begin{array}{r} 68 \\ - 29 \\ \hline 3F \end{array}$$

---

It is important to note that the octal and hexadecimals systems are used for human convenience only, and the computer system actually converts the octal and hex numbers into binary strings and operates on the binary digits.

## Binary Data Codes

Data codes translate information (alpha, numeric, or control characters) to a form that can be transferred electronically and then converted back to its original form. A code's efficiency is a measure of its ability to utilize the maximum capacity of the bits and to recover from error. In the evolution of the various codes, there has been a steady increase in their efficiency to transfer data. A brief discussion of the commonly used codes follows.

## Binary Code

It is possible to represent $2^n$ different symbols in a purely binary code of n bits. The binary code is a direct conversion of the decimal number to the binary. This is illustrated in Table 2-5. It is the most commonly used code in computers because it is a systematic arrangement of the digits.

It is also a weighted code, where each column has a magnitude of $2^n$ associated with it, and it is easy to translate. In Table 2-5, note that the least significant bit (LSB) alternates every time, whereas the second least significant bit repeats every two times, the third least significant bit repeats every four times, and so on.

## Baudot Code

The Baudot code was the first successful data communications code. It is also known as the International Telegraphic Alphabet #2 (ITA#2). The code was meant primarily for text transmission. It has only uppercase letters and is used with punched paper tape units on teletypewriters. It uses five consecutive bits and an additional start/stop bit to represent a data character as shown in Figure 2-2. It transfers asynchronous data at a very slow rate (10 characters per second) using a teletypewriter.

Most early teletypewriters used basically the same circuit as the telegraph with the mechanics of a typewriter. As with the telegraph, the

| Decimal | Binary | Decimal | Binary |
|---------|--------|---------|--------|
| 0 | 00000 | 12 | 01100 |
| 1 | 00001 | 13 | 01101 |
| 2 | 00010 | 14 | 01110 |
| 3 | 00011 | 15 | 01111 |
| 4 | 00100 | 16 | 10000 |
| 5 | 00101 | 17 | 10001 |
| 6 | 00110 | 18 | 10010 |
| 7 | 00111 | 19 | 10011 |
| 8 | 01000 | 20 | 10100 |
| 9 | 01001 | 21 | 10100 |
| 10 | 01010 | 22 | 10110 |
| 11 | 01011 | 23 | 10111 |

Table 2-5. Binary to Decimal Code.

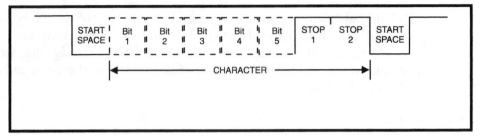

**Figure 2-2.  Baudot character communication format.**

teletypewriter had to have at one end a method of knowing when the other end wanted to transmit, so a mark signal (current) would be sent as a "line idle" signal. Since a mark is the idle condition, the first element or bit of any code would have to be a space (no current). This bit is known as the "start space" as shown in Figure 2-2. Also, the "current on" (mark) condition needs to exist after the character code pulses have been sent, so the receiver device will know when the character is complete and separate this transmission character from the next character to be transmitted. This period of current is know as the "stop mark" and is either 1, 1.42, or 2 elements in duration. The bit time or duration is determined by the teletype's motor speed.

The Baudot or teletypewriter code is given in Table 2-6. There are 26 uppercase letters, 10 numerals, and various items of punctuation and teletype control. This code uses five bits (2 to a 5th power) or 32 patterns. However, the code developers used the mechanical shift of the

| Character Case | | Bit Pattern | Character Case | | Bit Pattern |
|---|---|---|---|---|---|
| Lower | Upper | 54321 | Lower | Upper | 54321 |
| A | – | 00011 | Q | 1 | 10111 |
| B | ? | 11001 | R | 4 | 01010 |
| C | : | 01110 | S | ' | 00101 |
| D | $ | 01001 | T | 5 | 10000 |
| E | 3 | 00001 | U | 7 | 00111 |
| F | ! | 01101 | V | ; | 11110 |
| G | & | 11010 | W | 2 | 10011 |
| H | # | 10100 | X | / | 11101 |
| I | 8 | 00110 | Y | 6 | 10101 |
| J | Bell | 01011 | Z | " | 10001 |
| K | ( | 01111 | Letters Shift Down | | 11111 |
| L | ) | 10010 | Figures Shift Up | | 11011 |
| M | . | 11100 | Space | | 00100 |
| N | , | 01100 | Carriage Return | | 01000 |
| O | 9 | 11000 | Line Feed | | 00010 |
| P | 0 | 01101 | Blank or null | | 00000 |

**Table 2-6.  Baudot Code.**

teletypewriter to produce 26 patterns out of a possible 32 for letters and 26 patterns shifted for numbers and punctuation. Only 26 were available in either shift because 6 patterns were the same for both as shown in Table 2-6. The 6 common patterns are carriage return, line feed, shift up (figures), shift down (letters), space, and blank (no current).

This binary code is still the most efficient code for narrative text in terms of transmission overhead because it requires very little machine operation or error detection. While this code is no longer widely used, at one time it was the most extensively used binary transmission code.

A disadvantage of the code is that it can represent only the 58 characters shown in Table 2-6. Other limitations of Baudot code are the sequential nature of the code, the high overhead, and the lack of error detection.

## BCD Code

As computer and data communications technology improved, more efficient codes were developed. The BCD, binary coded decimal code was first used to perform internal numeric calculations within data processing devices. The BCD code is commonly used in programmable controllers to code data to numeric light-emitting diode (LED) displays and from panel-mounted digital thumb wheel units. Its main disadvantages are that it has no alpha characters and no error checking capability. A listing of the BCD code for decimal numbers from 0 to 19 is given in Table 2-7.

Note that four-bit groups are used to represent the decimal numbers 0 through 9. To represent higher numbers, such as 10 through 19 another four-bit group is used and place to the left of the first four-bit group.

| Decimal | BCD Code | Decimal | Binary |
|---------|----------|---------|--------|
| 0 | 0000 0000 | 10 | 0001 0000 |
| 1 | 0000 0001 | 11 | 0001 0001 |
| 2 | 0000 0010 | 12 | 0001 0010 |
| 3 | 0000 0011 | 13 | 0001 0011 |
| 4 | 0000 0100 | 14 | 0001 0100 |
| 5 | 0000 0101 | 15 | 0001 0101 |
| 6 | 0000 0110 | 16 | 0001 0110 |
| 7 | 0000 0111 | 17 | 0001 0111 |
| 8 | 0000 1000 | 18 | 0001 1000 |
| 9 | 0000 1001 | 19 | 0001 1001 |

Table 2-7.  BCD Code.

**EXAMPLE 2-11**

**Problem:** Convert the following decimal numbers to BCD:

   (a) 276,
   (b) 567,
   (c) 719, and
   (d) 4500.

**Solution:** Using Table 2-7, the decimal numbers can be expressed in BCD code as follows:

   (a) 276 = 0010 0111 0110,
   (b) 567 = 0101 0110 0111,
   (c) 719 = 0111 0001 1000, and
   (d) 4500 = 0100 0101 0000 0000

## ASCII Code

The most widely used code is ASCII, the American Standard Code for Information Interchange, which was developed in 1963. This code has 7 bits for data (allowing 128 characters) as shown in Table 2-8.

| Bits | 7<br>6<br>5 | 0<br>0<br>0 | 0<br>0<br>1 | 0<br>1<br>0 | 0<br>1<br>1 | 1<br>0<br>0 | 1<br>0<br>1 | 1<br>1<br>0 | 1<br>1<br>1 |
|---|---|---|---|---|---|---|---|---|---|
| 4321 | HEX | 0 | 1 | 2 | 3 | 4 | 5 | 6 | 7 |
| 0000 | 0 | NUL | DLE | SP | 0 | @ | P | ` | p |
| 0001 | 1 | SOH | DC1 | ! | 1 | A | Q | a | q |
| 0010 | 2 | STX | DC2 | " | 2 | B | R | b | r |
| 0011 | 3 | ETX | DC3 | # | 3 | C | S | c | s |
| 0100 | 4 | EOT | DC4 | $ | 4 | D | T | d | t |
| 0101 | 5 | ENQ | NAK | % | 5 | E | U | e | u |
| 0110 | 6 | ACK | SYN | & | 6 | F | V | f | v |
| 0111 | 7 | BEL | ETB | , | 7 | G | W | g | w |
| 1000 | 8 | BS | CAN | ( | 8 | H | X | h | x |
| 1001 | 9 | HT | EM | ) | 9 | I | Y | i | y |
| 1010 | A | LF | SUB | * | : | J | Z | j | z |
| 1011 | B | VT | ESC | + | ; | K | [ | k | { |
| 1100 | C | FF | FS | , | < | L | \ | l | \| |
| 1101 | D | CR | GS | – | = | M | ] | m | } |
| 1110 | E | SO | RS | . | > | N | ^ | n | ~ |
| 1111 | F | SI | US | / | ? | O | — | o | DEL |

Table 2-8. ASCII Code.

The ASCII code can operate synchronously or asynchronously with 1 or 2 stop bits. ASCII format has 32 control characters as shown in Table 2-9. These control codes are used to indicate, modify, or stop a control function in the transmitter or receiver. Seven of the ASCII control codes are called *format effectors*. They pertain to the control of a printing device. The use of format effectors increases code efficiency and speed by replacing frequently used character combinations with a single code. The following is a list of the format effectors used in the ASCII code: BS (backspace), HT (horizontal tab), LF (line feed), VT (vertical tab), FF (form feed), CR (carriage return), and SP (space).

A problem with using only 7-bit codes like ASCII occurs when computers are linked together. Most computers normally operate using 8-bit bytes or multiples of 8-bit bytes, such as 16 or 32 bits. The ASCII system uses 7 bits for coding and adds an eighth bit for parity checking. Parity means to count the number of ones in a character and to add a one or zero to the character string to produce an even or odd number of bits. If odd parity is chosen, then the parity bit will be the value required to ensure an odd number of ones in the character, including the parity bit. Since a parity bit is used the 256 combinations that are possible with an 8-bit string are reduced to the 128 characters of ASCII.

---

**EXAMPLE 2-12**

**Problem:** Express the words PUMP 100 ON using ASCII code. Use Hex notation for brevity.

**Solution:** Using Table 2-8, we obtain:

Message:       PUMP 100 ON
ASCII (Hex) String:       50 55 40 50 20 31 30 30 20 4F 4E

---

## EBCDIC Code

The extended binary coded decimal interchange code (EBCDIC) is an IBM proprietary data communications code. It uses all 8 bits to allow for up to 256 combinations as shown in Table 2-10. This code seemed a step backward in the development of data communication codes because it has no parity for error checking. However, other external error checking methods are used in the transmission of this code. Its main advantage is that its 8-bit code can accommodated up to 256 combinations.

| Mnemonic | Meaning | Mnemonic | Meaning |
|----------|---------|----------|---------|
| NUL | Null | DLE | Data Link Escape |
| SOH | Start of Heading | DC1 | Device Control 1 |
| STX | Start of Text | DC2 | Device Control 2 |
| ETX | End of Text | DC3 | Device Control 3 |
| EOT | End of Transmission | DC4 | Device Control 4 |
| ENQ | Enquiry | NAK | Negative Acknowledge |
| ACK | Acknowledge | SYN | Synchronous Idle |
| BEL | Bell | ETB | End of Transmission Block |
| BS | Backspace | CAN | Cancel |
| HT | Horizontal Tabulation | EM | End of Medium |
| LF | Line Feed | SUB | Substitute |
| VT | Vertical Tabulation | ESC | Escape |
| FF | Form Feed | FS | File Separator |
| CR | Carriage Return | GS | Group Separator |
| SO | Shift Out | RS | Record Separator |
| SI | Shift In | US | Unit Separator |
|  |  | DEL | Delete |

**Table 2-9.  Legend for ASCII Control Characters.**

There is a problem when we try to interface a personal computer (PC) to a mini or mainframe computer because PCs use ASCII coding while minis and mainframes use a different coding method—in IBM's case EBCDIC. This means that they cannot communicate directly without some form of translator, generally a software program or firmware in a protocol converter.

Table 2-10 is only a representative example. EBCDIC codes for punctuation and controls differ with the devices that they are intended to operate with.

| Bits | 8 7 6 5 | 0 0 0 0 | 0 0 0 1 | 0 0 1 0 | 0 0 1 1 | 0 1 0 0 | 0 1 0 1 | 0 1 1 0 | 0 1 1 1 |
|---|---|---|---|---|---|---|---|---|---|
| 4321 | HEX | 0 | 1 | 2 | 3 | 4 | 5 | 6 | 7 |
| 0000 | 0 | NUL | DLE | DS | | SP | & | – | |
| 0001 | 1 | SOH | DC1 | SOS | | | | / | |
| 0010 | 2 | STX | DC2 | FS | SYN | | | | |
| 0011 | 3 | ETX | DC3 | | | | | | |
| 0100 | 4 | PT | RES | SYP | PN | | | | |
| 0101 | 5 | HT | NL | LF | RS | | | | |
| 0110 | 6 | LC | BS | ETB | UC | | | | |
| 0111 | 7 | DEL | IL | ESC | EOT | | | | |
| 1000 | 8 | | CAN | | | | | | |
| 1001 | 9 | RLF | EM | | | | | | \ |
| 1010 | A | SMM | CC | SM | | @ | ! | \| | : |
| 1011 | B | VT | | | | . | $ | , | # |
| 1100 | C | FF | IFS | | DC4 | < | * | % | & |
| 1101 | D | CR | IGS | ENQ | NAK | ( | ) | — | ' |
| 1110 | E | SO | IRS | ACK | | + | ; | > | = |
| 1111 | F | SI | IUS | BEL | SUB | | – | ? | " |

| Bits | 8 7 6 5 | 1 0 0 0 | 1 0 0 1 | 1 0 1 0 | 1 0 1 1 | 1 1 0 0 | 1 1 0 1 | 1 1 1 0 | 1 1 1 1 |
|---|---|---|---|---|---|---|---|---|---|
| 4321 | HEX | 8 | 9 | A | B | C | D | E | F |
| 0000 | 0 | | | ~ | | { | } | \ | 0 |
| 0001 | 1 | a | j | | | A | J | | 1 |
| 0010 | 2 | b | k | s | | B | K | S | 2 |
| 0011 | 3 | c | l | t | | C | L | T | 3 |
| 0100 | 4 | d | m | u | | D | M | U | 4 |
| 0101 | 5 | e | n | v | | E | N | V | 5 |
| 0110 | 6 | f | o | w | | F | O | W | 6 |
| 0111 | 7 | g | p | x | | G | P | X | 7 |
| 1000 | 8 | h | q | y | | H | Q | Y | 8 |
| 1001 | 9 | i | r | z | | I | R | Z | 9 |

Table 2-10. EBCDIC Code.

## EXERCISES

2.1   Convert the binary number 10010 to a decimal number.

2.2   Find the binary equivalent to the decimal number 115.

2.3   Add the binary number 10101 to the binary number 1101 and find the decimal result.

2.4   Convert the following decimal numbers to their hexadecimal equivalents: (a) $656_{10}$, (b) $2566_{10}$, (c) $270_{10}$, (d) $29_{10}$, and (e) $12,056_{10}$.

2.5   Convert the following hexadecimal numbers to their decimal equivalents: (a) $57_{16}$, (b) $754_{16}$, (c) $345_{16}$, (d) $5687_{16}$, and (e) $455_{16}$.

2.6   Convert the following binary numbers to their hexadecimal equivalents: (a) $101_2$, (b) $1011111_2$, (c) $1011100_2$, (d) $01101_2$ and (e) $11001111101_2$.

2.7   Add the following hexadecimal numbers: (a) A34 + E45, (b) C87 + AA1, (c) 34 + CCE, (d) 397 + 1A2, and (e) 3D4 + 167.

2.8   Perform the following subtractions of the following hexadecimal numbers: (a) B8—25, (b) 36—A, (c) lC1—FC, (d) 300—3, and (e) 2BE—234.

2.9   Write the ASCII binary code for the message "Temperature in Furnace is High" using binary and Hex strings.

2.10  Write the EBCDIC binary code for the message "Level in Evaporator is Low" using binary and Hex strings.

## BIBLIOGRAPHY

1.   Kintner, P. M., *Electronic Digital Techniques*, McGraw-Hill Book Company, 1968.

2.   Floyd, T. L., *Digital Logic Fundamentals*, Charles E. Merrill Publishing Company, 1977.

3.   Malvino, A. P., *Digital Computer Electronics — An Introduction to Microcomputers*, Second Edition, McGraw-Hill Book Company, 1983.

4.   Wickers, W. E., *Logic Design with Integrated Circuits*, John Wiley & Sons, Inc., 1968.

5.   Thompson, L. M., *Industrial Data Communications Fundamentals and Applications*, ISA, 1991.

# 3

# Digital Logic Fundamentals

## Introduction

The application of digital techniques to process control requires an understanding of digital system fundamentals. A working knowledge of digital logic principles is also required in the design and implementation of logic systems and microcomputer-based control systems. In this chapter, the basic concepts of logic functions and logic design will be discussed. We will also discuss relay ladder logic design.

## Logic Functions

In binary logic control systems, binary numbers 1 and 0 are represented by voltage levels, relays contact status, switch position, etc. For example, in transistor-transistor logic (TTL) gates, a binary 1 is represented by a voltage signal in the range of 2.4 to 5.0 volts, and a binary 0 is represented by a voltage level between 0 and 0.8 volt. Solid-state electronic circuits are available that can be used to manipulate digital signals to perform a variety of logical functions, such as NOT, AND, OR, NAND, NOR, and NOT. In hardwired electrical logic systems, relays are used to implement the logic functions.

### NOT Function

An important logic function is the NOT or inversion function. The NOT, or logic inverter, produces an output opposite to the input. An inversion bar is drawn over a logic variable to indicate the NOT function. For example, if a NOT operation is performed on a logic variable $A$, it is designated by $\overline{A}$. Table 3-1 lists the results of the NOT function.

| Input | Output |
|-------|--------|
| A | $Z = \overline{A}$ |
| 0 | 1 |
| 1 | 0 |

Table 3-1. NOT Function Truth Table.

## OR Function

An OR function, with two or more inputs and a single output, operates in accordance with the following definition: *The output of an OR function assumes the 1 state if one or more inputs assume the 1 state.*

The inputs to a logic function or gate are designated by $A$, $B$, . . ., $N$ and the output by $Z$. It is assumed that the inputs and outputs can take one of two possible values, either 0 or 1. The logic expression for this function is $Z = A + B + \ldots + N$. A truth table, which contains a tabulation of all possible input values and their corresponding outputs, is also given in Table 3-2 for a two-input OR function.

An example of OR logic in process control would be as follows: If the water level in a hot water heater is low, OR the temperature in the tank is too high, a logic system can be designed to turn off the heater in the system using logic circuits or relays.

The following logic identities for OR functions can be easily verified using the two-input truth table for the OR function given in Table 3-2.

$$A + B + C = (A + B) + C = A + (B + C) \tag{3-1}$$

$$A + B = B + A \tag{3-2}$$

$$A + A = A \tag{3-3}$$

$$A + 1 = 1 \tag{3-4}$$

| Inputs | | Output |
|--------|---|--------|
| A | B | Z |
| 0 | 0 | 0 |
| 0 | 1 | 1 |
| 1 | 0 | 1 |
| 1 | 1 | 1 |

Table 3-2. Two Input OR Function Truth Table.

$$A + 0 = A \qquad (3\text{-}5)$$

Remember that $A$, $B$, and $C$ can take on only the value of 0 or 1.

## AND Function

An AND function has two or more inputs and a single output, and it operates in accordance with the following rule: *The output of an AND gate assumes the 1 state if and only if all the inputs assume the 1 state.* The general equation for the AND function is given by $ABC \ldots N = Z$. A two-input AND function truth table is given by Table 3-3.

| Inputs | | Output |
|:---:|:---:|:---:|
| A | B | Z |
| 0 | 0 | 0 |
| 0 | 1 | 0 |
| 1 | 0 | 0 |
| 1 | 1 | 1 |

**Table 3-3. Two-Input AND Function Truth Table.**

The logic expressions for the AND function are as follows:

$$ABC = (AB)C = A(BC) \qquad (3\text{-}6)$$

$$AB = BA \qquad (3\text{-}7)$$

$$AA = A \qquad (3\text{-}8)$$

$$A1 = A \qquad (3\text{-}9)$$

$$A0 = 0 \qquad (3\text{-}10)$$

These identities can be verified by reference to the definitions of the AND gate and by using a truth table for the AND gate.

For example, Equation 3-11 ($A1 = A$) can be verified. Let $B = 1$ and tabulate $AB = Z$ in a truth table.

| Inputs | | Output |
|:---:|:---:|:---:|
| A | B | Z |
| 0 | 1 | 0 |
| 1 | 1 | 1 |

Note that $A = Z$, so that $A1 = A$ for both values (0 and 1) of $A$.

Some important auxiliary identities used in logic design are as follows:

$$A + AB = A \tag{3-11}$$

$$A + \overline{A}B = A + B \tag{3-12}$$

$$(A + B)(A + C) = A + BC \tag{3-13}$$

These identities are important because they help reduce the number of logic elements or gates required to implement a logic function.

## NOR Function

Another logic gate in common use is the NOR gate, and Table 3-4 shows its operation for two inputs. It produces a logic 1 result if and only if all inputs are logic 0. Notice that the output of the NOR function is opposite from the output of an OR gate.

| Inputs | | Output |
|---|---|---|
| A | B | Z |
| 0 | 0 | 1 |
| 0 | 1 | 0 |
| 1 | 0 | 0 |
| 1 | 1 | 0 |

Table 3-4. Two Input NOR Function Truth Table.

## NAND Function

Another logic operation of interest is the NAND gate; its operations are summarized in Table 3-5 for a two-input NAND function. Note that the output of the NAND function is the exact opposite of the AND function output. When both inputs to the NAND are 1, the output is 0. In all other configurations, the NAND function output is 1.

| Inputs | | Output |
|---|---|---|
| A | B | Z |
| 0 | 0 | 1 |
| 0 | 1 | 1 |
| 1 | 0 | 1 |
| 1 | 1 | 0 |

Table 3-5. Two Input NAND Function Truth Table.

## DeMorgan's Theorem

A great deal of logic design is based upon a set of rules called DeMorgan's Theorem. This theorem demonstrates that any logic equation can be synthesized from AND, NOT, and NAND gates or from OR, NOT, and NOR gates. DeMorgan's theorem can be summarized: *If the inversion bar (NOT function) is broken between two logic variables, the logic symbol connecting the variable can be changed.*

The reverse is also true: *If the inversion bar is joined between two variables, the logic symbol connecting the variables can be changed.*

In equation form, DeMorgan's Theorem is:

$$\overline{(AB \cdots N)} = \overline{A} + \overline{B} + \cdots + \overline{N} \qquad (3\text{-}14)$$

$$\overline{(A + B + \cdots + N)} = \overline{A}\,\overline{B} \cdots \overline{N} \qquad (3\text{-}15)$$

DeMorgan's Laws are frequently used to simplify inverted logic equations or to convert a logic expression into a different form or type. For example, according to DeMorgan's Laws $\overline{(AB)} = \overline{A} + \overline{(B}$ and $\overline{(A + B)} = A\overline{B}$.

DeMorgan's Theorem and some other basic logic laws and identities have been summarized in Table 3-6 for easy reference. Control logic equations can vary from very simple to complex, but they always satisfy the basic logic rules listed in Table 3-6. These rules are a result of simple combination of the basic truth tables and they can be used to simplify logic equations and circuits.

These laws are very useful in logic design, especially when used in conjunction with an operation called involution. This operation consists of taking the NOT or inversion of a logic variable twice or $\overline{\overline{A}} = A$.

| Commutative | $A + B = B + A$ <br> $AB = BA$ |
|---|---|
| Associative | $A + (B + C) = (A + B) + C$ <br> $A + BC = (A + B)(A + C)$ |
| Distributive | $A(B + C)( = AB + AC$ <br> $A + BC = (A + B)(A + C)$ |
| Absorption | $A + AB = A$ |
| DeMorgan | $\overline{AB)} = \overline{A} + \overline{B}$ <br> $\overline{A + B)} = \overline{A}\overline{B}$ |

**Table 3-6. Logic Rules.**

Involution and DeMorgan's Theorem work together to interchange AND and OR functions.

In applying these laws to logic design, the least complex logic realization is found by factoring out all common terms. The application of these rules to logic circuit design can be demonstrated by the following examples.

---

**EXAMPLE 3-1**

**Problem:** Reduce the logic expression:

$$Z = ABC + AB\overline{C} + \overline{A}BC + \overline{A}\overline{B}C$$

to a simpler form.

**Solution:** A simpler expression can be obtained by using logic laws and identities as follows:
First, using the associative law to collect common terms, we obtain

$$Z = ABC + AB\overline{C} + \overline{A}CB + \overline{A}C\overline{B}$$

Next, using the distribution law, we find that

$$Z = AB(C + \overline{C}) + \overline{A}C(B + \overline{B})$$

Finally, we use the identities $C + \overline{C} = 1$ and $B + \overline{B} = 1$ to obtain the simplified expression,

$$Z = AB + \overline{A}C$$

---

## Logic Function Symbols

There are two common sets of symbols used in process control applications to represent logic function: graphic and ladder. The graphic symbols are generally used on engineering drawings to convey the overall logic plan for a discrete or batch control system and they are based on ANSI/ISA S5.2 Binary Logic Diagrams for Process Operations. The ladder logic symbols are used to describe a logic control plan if relays are being used to implement the control system or if programmable controller ladder logic is used to implement the control plan. Figure 3-1 compares these two sets of symbols for the most common logic functions encountered in logic control design.

## Ladder Logic Diagrams

Ladder diagrams are a traditional method used to describe electrical logic controls. These circuits are called ladder diagrams because they look like ladders with rungs. Each rung of a ladder is numbered so that we can easily cross reference between sections on the drawing that describe the control.

We can appreciate the utility of ladder logic diagrams by investigating a simple control example. Figure 3-2 shows a typical process level control application. In this application, let us assume that the flow into the tank is random and we need to control the level in the tank by opening or closing the on/off electric solenoid valve (LV-1) based on the level sensed in the tank by a level switch (LSH-1). We will also provide the operator with a three position hand, off and automatic (HOA) switch to manually turn the valve on or off or the option to select automatic control using the level switch to maintain the proper level in the tank.

The ladder logic design for this application is shown in Figure 3-3. If the HOA switch is in the automatic position and the level switch is closed, the solenoid will be energized. This is a simple example of a logical AND function in process control. If the HOA switch is in the hand or manual position, the valve will also be turned on. If we designate the logic variables as follows—*A* for hand position, *B* for

| LOGIC FUNCTION | GRAPHIC SYMBOL | LADDER SYMBOL |
|---|---|---|
| NOT, $Z = \overline{A}$ | $A \multimap Z = \overline{A}$ | |
| OR, $Z = A + B$ | | |
| AND, $Z = AB$ | | |

**Figure 3-1. Comparison of logic symbols.**

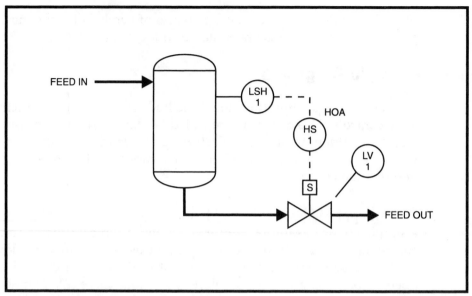

**Figure 3-2. On-off control of tank level.**

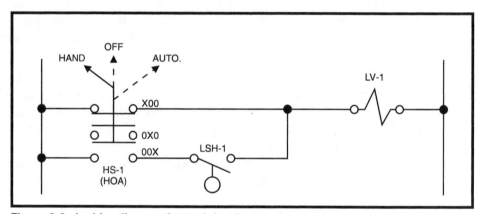

**Figure 3-3. Ladder diagram for tank level control.**

automatic position, $C$ for level switch LSH-1, and $Z$ for the solenoid valve LV-1—then the logic equation for the control system is $Z = A + BC$.

A more complex application might be to control tank level between two level switches, a level switch high (LSH) and a level switch low (LSL). In this application, a pump supplying fluid to a tank is turned on and off to maintain the liquid level in the tank between the two level switches. The process is shown in Figure 3-4 and it uses steam at a regulated flow to boil down a liquid to produce a more concentrated solution, which is drained off periodically by opening a manual valve on the bottom of the tank. Note that a small diamond symbol is used to indicate that the pump is interlocked with the level switches on the tank.

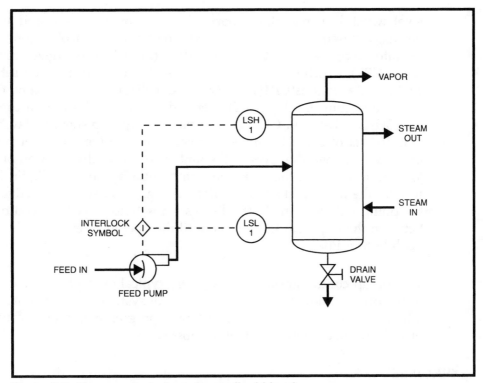

**Figure 3-4. Pump control of condenser liquid level.**

The electrical ladder diagram used to control the feed pump and hence the liquid level in the process tank is shown in Figure 3-5. To explain the logic of the control system, we will assume the tank is empty, the low level switch is closed, and the high level switch is closed. When a

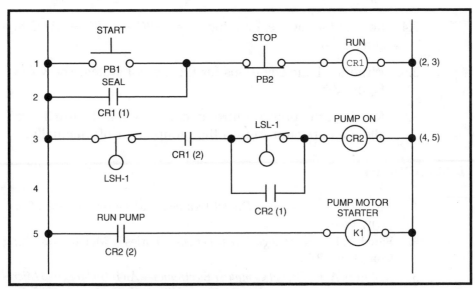

**Figure 3-5. Ladder diagram for pump control.**

level switch is activated, the normally open contacts are closed and the normally closed contacts are opened. To start the control system, the operator depresses the start push button (PB1). This energizes control relay (CR1), which seals in the start push button with the first set of contacts, denoted as CR1(1) on the ladder diagram. At the same time, the second control relay (CR2) is energized in rung 3 through contacts on LSH-1, LSL-1, and CR1. This turns on the pump starter relay K1. The first set of contacts on CR2 is used to seal in the low level switch contacts, so when the level in the tank rises above the position of the low level switch on the tank, the pump will stay until the liquid level reaches the high level switch. After the high level switch is activated, the pump will be turned off. The system will now cycle on and off between the high and low levels until the operator depresses the stop push button PB-2.

This pump control application is a very typical example of logic control in the process industries. The logic control system is implemented using a hard-wired relay based logic system, a programmable logic controller, or a distributed control system in most cases.

## EXERCISES

3.1. Verify the logic identity $A + B = B + A$ using truth tables.

3.2. Verify the logic identity $A + AB = A$ using truth tables

3.3. Verify the logic identity $A + 1 = 1$ using an AND truth table.

3.4. Reduce the logic expression: $Z = AB\overline{C} + \overline{A}BC + \overline{A}\,\overline{B}C$ to a simpler form.

3.5. Write the logic equations for the control system shown in Figure 3-5.

3.6. Redesign the pump control circuit shown in Figure 3-5 to allow an operator to turn the pump on or off manually.

## BIBLIOGRAPHY

1. Kintner, P. M., *Electronic Digital Techniques*, McGraw-Hill Book Company, 1968.
2. Floyd, T. L., *Digital Logic Fundamentals*, Charles E. Merrill Publishing Company, 1977.
3. Malvino, A. P., *Digital Computer Electronics — An Introduction to Microcomputers*, Second Edition, McGraw-Hill Book Company, 1983.

4.  Grob, B., *Basic Electronics*, Fifth Edition, McGraw-Hill Book Company, 1984.

5.  Budak, A., *Passive and Active Network Analysis and Synthesis*, Houghton Mufflin Co., 1974.

6.  Boylestad, R. L. and Nashelsky, L., *Electronic Devices and Circuit Theory*, Third Edition, Prentice-Hall, Inc., 1982.

7.  Clare, C. R., *Designing Logic Systems Using State Machines*, McGraw-Hill Book Company, 1973.

# 4

# Electrical and Electronic Fundamentals

## Introduction

The design of programmable controller systems requires knowledge of the basic principles of electricity and electronics. This chapter will discuss the fundamentals of electricity and then investigate the electrical and electronic circuits that are commonly encountered in the instrumentation and control field. The operation and purpose of electrical and electronic devices such as power supplies, relays, solenoid valves, contactors, and control switches and indicating lights will be discussed.

## Fundamentals of Electricity

Electricity is a fundamental force in nature that can produce heat, motion, light, and many other physical effects. This force is an attraction or repulsion between electric charges called electrons and protons. An electron is a small atomic particle having a negative electric charge. A proton is a basic atomic particle with a positive charge. It is the arrangement of electrons and protons that determines the electrical characteristics of all substances. As an example, this page of paper has electrons and protons in it, but there is no evidence of electricity because the number of electrons equals the number of protons. In this case, the opposite electrical charges cancel, making the paper electrically neutral.

If we want to use the electrical forces associated with positive and negative charges, work must be done to separate the electrons and protons. For example, an electric battery can do such work because its chemical energy separates electrical charges to produce an excess of electrons at its negative terminal and an excess of protons at its positive terminal.

The basic terms encountered in electricity are electric charge ($Q$), current ($I$), voltage ($V$), and resistance ($R$). A charge of a very large number of electrons and protons is required for common applications of electricity. Therefore, it was convenient to define a practical unit called the coulomb ($C$) as equal to the charge of $6.25 \times 10^{18}$ electrons or protons. The symbol for electric charge is $Q$, standing for quantity of charge, and a charge of $6.25 \times 10^{18}$ electrons or protons is stated as $Q = 1\,C$. Voltage is the potential difference force produced by differences in opposite electrical charges. Fundamentally, the volt is a measure of the work required to move an electric charge. When one joule of work is required to move one coulomb between two points, the potential difference is 1 volt ($V$).

When the potential difference between two different charges forces a third charge to move, the charge in motion is called electric current. If the charges move at the rate of one coulomb per second past a given point, the amount of current is defined as one ampere ($A$). In equation form, $I = dQ/dt$, where $I$ is the instantaneous current in amperes and $dQ$ is the differential amount charge in coulombs passing a given point during the time period ($dt$) in seconds. If the current flow is constant, it is simply given by $I = Q/t$, where $Q$ is the amount of current flowing past a given point in $t$ seconds.

---

**EXAMPLE 4-1**

**Problem:** A steady flow of 12 coulombs of charge passes a given point in a copper conductor every 3 seconds. What is the current flow?

**Solution:** Since constant current flow is defined as the amount of charge ($Q$) that passes a given point per period of time ($t$) in seconds, we have

$$I = Q/t = 12\ C/3\ s$$

$$I = 4\ amp$$

---

The fact that a conductor carrying electric current can become hot is evidence that the work done by the applied voltage in producing current must be meeting some form of opposition. This opposition, which limits current flow, is called resistance.

## Conductivity, Resistivity, and Ohm's Law

An important physical property of some material is called conductivity, i.e., the ability to pass electric current. Suppose we have an electric wire (conductor) of length $L$ and cross-sectional area $A$, and we apply a voltage $V$ between the ends of the wire. If $V$ is in volts and $L$ is in meters, we can define the voltage gradient $(E)$, as

$$E = \frac{V}{L} \text{ (volt/meter)} \tag{4-1}$$

Now, if a current I in amperes flows through a wire of area $A$ in meters squared $(m^2)$, we can define the current density $J$ as

$$J = \frac{I}{A} \text{ (amps/meters}^2\text{)} \tag{4-2}$$

The conductivity $C$ is defined as the current density per unit voltage gradient $E$ or, in equation form, we have

$$C = \frac{J}{E} \text{ (amps/meter}^2\text{)/(volts/meter)}$$

$$C = \frac{I/A}{V/L} \tag{4-3}$$

or

Resistivity $(r)$ is defined as the inverse of conductivity, or

$$r = \frac{1}{C} \tag{4-4}$$

The fact that resistivity is a natural property of certain materials leads to the basic principle of current flow called Ohm's Law.

Consider a wire of length $L$ and area $A$. If it has resistivity, $r$, then its resistance $R$ is

$$R = r\frac{L}{A} \tag{4-5}$$

The unit of resistance is the ohm and it is denoted by the Greek letter omega, $\Omega$. Since the resistivity, $r$, is the reciprocal of conductivity, we obtain the following:

$$r = \frac{V/L}{I/A} \tag{4-6}$$

When Equation 4-6 is substituted into Equation 4-5, we obtain

$$R = \frac{V}{I} \qquad\qquad (4\text{-}7)$$

This relationship, $R = V/I$ or $V = IR$, is called Ohm's Law, which assumes that the resistance of the material used to carry the current flow will have a linear relationship between the voltage applied and the current flow, i.e., if the voltage across the resistance is doubled, the current through it also doubles. The resistance of materials like carbon, aluminum, copper, silver, gold, and iron is linear and follows Ohm's Law. Carbon is the most common material used to manufacture a device with fixed resistance. This device is called a *resistor*.

## Wire Resistance

To compare the resistance and size of one conductor with another, a standard or unit size of conductor was established. A convenient unit of measurement for the diameter of a circular wire is the mil (0.001 inch) and a convenient unit of wire length is the foot. The standard unit of wire size in most cases is the *mil-foot*; that is, a wire is said to have a unit size if it has a diameter of 1 mil and a length of 1 foot.

The *circular mil* is the standard unit of wire cross-sectional area used in US wire sizing tables. Because the diameter of circular wire is normally only a small fraction of an inch, it is convenient to express these diameters in mils, to avoid the use of decimals. For example, the diameter of a 0.010-inch conductor is expressed as 10 mils instead of 0.010 inches. The circular mil area is defined as the square of the mil diameter of a conductor, or area (Cmil) $= d^2(\text{mil}^2)$. An example will be used to show how to calculate conductor area in circular mils.

---

**EXAMPLE 4-2**

**Problem:** Calculate the area in circular mils of a conductor with a diameter of 0.002 inch.

**Solution:** First, we must convert the diameter in inches to a diameter in mils. Since we defined 0.001 in. = 1 mil, this implies that 0.002 in. = 2 mils, so the circular mil area is $(2 \text{ mil})^2$ or 4 cmil.

---

A *circular-mil per foot* is a unit conductor 1 foot in length having a cross-sectional area of 1 circular mil. The cmil per foot is useful in making comparisons of the resistivity of various conductors. The specific resistance ($r$) in cmil-ohms per foot of some common solid metals at 20°C is given in Table 4-1.

Conductors used to carry electric current are normally manufactured of copper because of its low resistance and relatively low cost. We can use the specific resistance ($r$) given in Table 4-1 to find the resistance of a conductor of length ($L$) in feet and cross-sectional area ($A$) in cmil by using Equation 4-5, $R = rL/A$.

---

**EXAMPLE 4-3**

**Problem:** Find the resistance of 1000 feet of copper (drawn) wire having a cross-sectional area of 10,370 cmil, and whose wire temperature is 20°C.

**Solution:** The specific resistance of copper wire from Table 4-1 is 10.37 cmil-ohms/ft. Substituting the known values in Equation 4-5, the resistance is determined as follows:

$$R = r\frac{L}{A}$$

$$R = (10.37\text{cmil-}\Omega\text{/ft})(1000 \text{ ft})/(10,370 \text{ cmil})$$

$$R = 1\Omega$$

---

The equation for the resistance of a conductor, $R = rL/A$, can be used in many applications. For example, if $R$, $r$, and $A$ are known, the length can be determined by a simple mathematical transformation of the equation for the resistance of a conductor. A typical application is locating a problem ground point in a telephone line. Specially designed test equipment is used by the telephone company to find ground faults. This equipment operates on the principle that the resistance of a conductor varies directly with length such that the distance between a test point and a fault can be measured accurately with properly designed equipment.

| Material | r, cmil-$\Omega$/ft |
|---|---|
| Silver | 9.8 |
| Copper (drawn) | 10.37 |
| Gold | 14.70 |
| Aluminum | 17.02 |
| Tungsten | 33.20 |
| Brass | 42.10 |
| Steel | 95.80 |

Table 4-1. Specific resistance (*r*) at 20°C.

---

**EXAMPLE 4-4**

**Problem:** The resistance to ground on a faulty underground telephone line is 5 Ω. Calculate the distance to the point where the wire is shorted to ground, if the line is a copper conductor with a area of 1020 cmil and the ambient temperature of the conductor is 20°C.

**Solution:** To calculate the distance to the point where the wire is shorted to ground, we use $L = RA/r$. Since $R$ = 5 Ω, $A$ = 1,020 cmils, and $r$ = 10.37 cmil-Ω/ft, we have

$$L = RA/r$$

$$L = (5 \text{ Ω}) (1020 \text{ cmils})/10.37 \text{ cmil-Ω/ft}$$

$$L = 492 \text{ feet}$$

---

## Wire Gage Sizes

Electrical conductors are manufactured in sizes numbered according to a system known as the American Wire Gage (AWG). As can be seen in Table 4-2, the wire diameters become smaller as the gage numbers increase and the resistance per 1000 feet increases as the wire diameter decreases. The largest wire size listed in the table is 4 and the smallest is 24. This is the normal range of wire sizes typically encountered in process control applications. The complete AWG table goes from 0000 to 40; the larger and smaller sizes not listed in Table 4-2 are manufactured but not commonly encountered in process control.

---

**EXAMPLE 4-5**

**Problem:** Determine the resistance of 2500 feet of 14 AWG copper wire. Assume the wire temperature is 25°C.

**Solution:** Using Table 4-2, we see that 14 AWG wire has a resistance of 2.58 Ω per 1000 ft at 25°C. So the resistance of 2500 ft is calculated as follows:

$$R = (2.58 \text{ Ω}/1000 \text{ ft})(2500 \text{ ft}) = 6.5 \text{ Ω}$$

---

## Direct and Alternating Current

Basically, two types of voltage signals are encountered in process control and measurement: direct current (dc) and alternating current (ac). In direct current, the flow of charges is in just one direction. A battery is one example of a dc power source. A graph of a dc voltage

| AWG Number | Diameter, mil | Area cmil | Ω/1000 ft at 25°C | Ω/1000 ft at 65°C |
|---|---|---|---|---|
| 04 | 204 | 41700 | 0.253 | 0.292 |
| 06 | 162 | 26300 | 0.403 | 0.465 |
| 08 | 128 | 16500 | 0.641 | 0.739 |
| 10 | 102 | 10400 | 1.020 | 1.180 |
| 12 | 81.0 | 6530 | 1.620 | 1.870 |
| 14 | 64.0 | 4110 | 2.58 | 2.97 |
| 16 | 51.0 | 2580 | 4.09 | 4.73 |
| 18 | 40.0 | 1620 | 6.51 | 7.51 |
| 20 | 32.0 | 1020 | 10.4 | 11.9 |
| 22 | 25.3 | 642 | 16.5 | 19.0 |
| 24 | 20.1 | 404 | 26.2 | 30.2 |

Table 4-2. American Wire Gage for Solid Copper Wire.

versus time is shown in Figure 4-1a, and the symbol for a battery is shown in Figure 4-1b.

An alternating voltage source periodically reverses its polarity. Therefore, the resulting current flow in a closed circuit will reverse direction periodically. Figure 4-2a shows a sine-wave example of an ac voltage signal. The schematic symbol for an ac power source is shown in Figure 4-2b.

The 60 cycle ac power used in homes and industry is the most common example of ac power. The term "60 cycle" means that the voltage polarity and current direction go through 60 reversals or changes per second and the signal is said to have a frequency of 60 cycles per second. The unit for 1 cycle per second (cps) is called 1 hertz (Hz). Therefore, a 60 cycle per second signal has a frequency of 60 Hz.

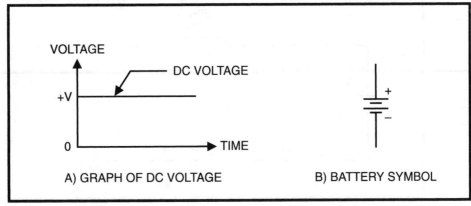

Figure 4-1. Steady dc voltage.

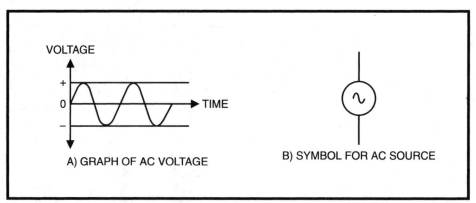

**Figure 4-2. Sine-wave ac voltage.**

## Series Resistance Circuits

When the components in a circuit are connected in successive order with the end of a component joined to the end of the next element, they form a series circuit. An example of a series circuit is shown in Figure 4-3.

In this circuit, the current ($I$) flows from the negative terminal of the battery through the two resistors ($R_1$ and $R_2$) and back to the positive terminal. According to Ohm's Law, the amount of current ($I$) flowing between two points in a circuit equals the potential difference ($V$) divided by the resistance ($R$) between these points. If $V_1$ is the voltage drop across $R_1$, $V_2$ is defined as the voltage drop across $R_2$, and the current ($I$) flows through both $R_1$ and $R_2$, we obtain from Ohm's Law

$$V_1 = IR_1 \text{ and } V_2 = IR_2$$

so that

$$V_t = IR_1 + IR_2$$

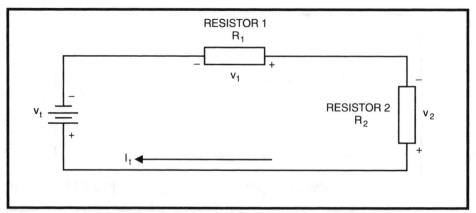

**Figure 4-3. Series resistance circuit.**

If we divided both sides of this equation by the current, $I$, in the series circuit, we obtain

$$V_t/I = R_1 + R_2$$

Since the total resistance of the series circuit is defined by Ohm's Law as the total voltage applied divided by the current in the circuit or $R_t = V_t/I$, the total resistance of the circuit is given by

$$R_t = R_1 + R_2$$

We can derive a more general equation for any number of resistors in series by using the classical Law of Conservation of Energy. According to this law, the energy or power supplied to a series circuit must equal the power dissipated in the resistors in the circuit.

Thus,

$$P_t = P_1 + P_2 + P_3 \cdots P_n$$

since power in a resistive circuit is given by $P = I^2R$, we obtain,

$$I^2R_t = I^2R_1 + I^2R_2 + I^2R_3 + \cdots + I^2R_n$$

If we divide this equation by $I^2$, we obtain the series resistance formula for a circuit with n resistors

$$R_t = R_1 + R_2 + R_3 + \cdots + R_n \tag{4-8}$$

---

**EXAMPLE 4-6**

**Problem:** Assume that the battery voltage in the series circuit of Figure 4-2 is 6 V dc and that resistor $R_1 = 1$ k$\Omega$ and resistor $R_2 = 2$ k$\Omega$. Find the total current flow ($I_t$) in the circuit and the voltage across $R_1$ and $R_2$.

**Solution:** To calculate the circuit current $I_t$, first find the total circuit resistance $R_t$ using Equation 3-7:

$$R_t = R_1 + R_2$$
$$R_t = 1 \text{ k}\Omega + 2 \text{ k}\Omega$$
$$R_t = 3 \text{ k}\Omega$$

*(continued on next page)*

**EXAMPLE 4-6, cont.**

Now, according to Ohm's Law:

$$I_t = \frac{V_t}{R_t}$$

$$I_t = \frac{6V}{3\ k\Omega} = 2\ mA$$

The voltage across $R_1$ (i.e., $V_1$) is given by

$$V_1 = R_1 I_t$$
$$V_1 = (1\ k\Omega)(2\ mA) = 2V$$

The voltage across $R_2$ (i.e., $V_2$) is obtained as follows:

$$V_2 = R_2 I_t$$
$$V_2 = (2\ k\Omega)(2\ mA) = 4\ V$$

## Parallel Resistance Circuits

When two or more components are connected across a power source, they form a parallel circuit. Each parallel path is called a branch, and each has its own current. In other words, parallel circuits have one common voltage across all the branches, but individual branch currents. These characteristics are in contrast to series circuits that have one common current, but individual voltage drops.

A parallel resistance circuit with two resistors across a battery is shown in Figure 4-4. The total resistance ($R_t$) across the power supply can be found by Ohm's Law by dividing the voltage across the parallel

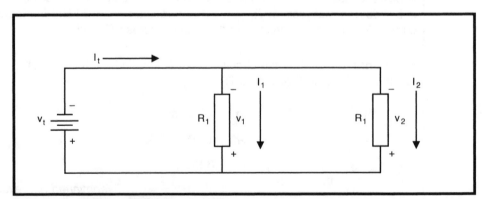

**Figure 4-4. Parallel resistance circuit.**

resistance by the total current of all the branches. In the circuit of Figure 4-4, $I_t = I_1 + I_2$ and $I_1 = V_t/R_1$, $I_2 = V_t/R_2$, so that $I_t = V_t/R_1 + V_t/R_2$. Since $I_t = V_t/R_t$, we obtain

$$\frac{1}{R_t} = \frac{1}{R_1} + \frac{1}{R_2}$$

or

$$R_t = \frac{R_1 \times R_2}{R_1 + R_2}$$

(4-9)

We can derive the general reciprocal resistance formula for any number of resistors in parallel from the fact that the total current $I_t$ is the sum of all of the branch currents, or

$$I_t = I_1 + I_2 + I_3 + \cdots + I_n$$

Since current flow is defined as $I = dQ/dt$, this is simply the law of electrical charge conservation:

$$\frac{dQ_t}{dt} = \frac{dQ_q}{dt} + \frac{dQ_2}{dt} + \frac{dQ_3}{dt} + \cdots + \frac{dQ_n}{dt}$$

This law implies that, since no charge accumulates at any point in the circuit, the differential charge, $dQ$, from the power source in the differential time period, $dt$, must appear as charges $dQ_1$, $dQ_2$, $dQ_3 \ldots$ $dQ_n$ through the resistors $R_1$, $R_2$, $R_3 \ldots R_n$ at the same time. Since the voltage across each branch is the applied voltage $V_t$ and $I = V/R$, we obtain,

$$\frac{V_t}{R_t} = \frac{V_t}{R_1} + \frac{V_t}{R_2} + \frac{V_t}{R_3} + \cdots + \frac{V_t}{R_n}$$

If we multiply both sides of this equation by $1/V_t$, we obtain the total resistance of n resistors in parallel, as follows:

$$\frac{1}{R_t} = \frac{1}{R_1} + \frac{1}{R_2} + \frac{1}{R_3} + \cdots + \frac{1}{R_n}$$

(4-10)

To illustrate the basic concepts of a parallel circuit, let's look at an example problem.

**EXAMPLE 4-7**

**Problem:** Assume that the voltage for the parallel circuit shown in Figure 4-4 is 12 *V* and we need to find the currents $I_t$, $I_1$, and $I_2$ and the parallel resistance $R_t$, given that $R_1 = 30$ k$\Omega$ and $R_2 = 30$ k$\Omega$

**Solution:** We can find the parallel circuit resistance $R_t$ by using Equation 4-9:

$$R_t = \frac{R_1 \times R_2}{R_1 + R_2}$$

$$R_t = \frac{(30\ k)\ (30\ k)}{30\ K + 30K}\Omega = 15\ \text{k}\Omega$$

The total current flow is given by

$$I_t = \frac{V_t}{R_t} = \frac{12\ V}{15\ \text{k}\Omega} = 0.8\ \text{mA}$$

The current in branch 1 ($I_1$) is obtained as follows:

$$I_1 = \frac{V_t}{R_1} = \frac{12\ V}{30\ \text{k}\Omega} = 0.4\ \text{mA}$$

Since $I_t = I_1 + I_2$, the current flow in branch 2 is given by

$$I_2 = I_t - I_1 = 0.8\ \text{mA} - 0.4\ \text{mA} = 0.4\ \text{mA}$$

## Wheatstone Bridge Circuit

The Wheatstone bridge circuit, named for English physicist and inventor Sir Charles Wheatstone (1802–1875), was one of the first electrical measuring instruments invented to accurately measure resistance. The bridge circuit is shown in Figure 4-5. The circuit has two parallel resistance branches with two series resistors in each branch and a galvanometer (*G*) connected across the branches. The galvanometer is a very sensitive electric charge measuring instrument that is momentarily connected across the bridge between points a and b. If these points are at the same potential, the meter will not deflect. The purpose of the circuit is to have the voltage drops balanced across the two parallel branches to obtain zero volts across the meter. In the Wheatstone bridge, the unknown resistance $R_x$ is balanced against a standard accurate resistor $R_s$ for precise measurement of resistance.

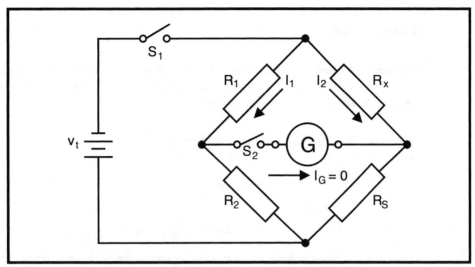

**Figure 4-5. Wheatstone bridge circuit.**

In the circuit shown, the switch $S_1$ is closed to apply the voltage $V_t$ to the four resistors in the bridge. To balance the circuit, the value of $R_x$ is varied until a zero reading is obtained on the meter with switch $S_2$ closed.

A typical application of the Wheatstone bridge circuit is to replace the unknown resistor $R_x$ with a temperature bulb. A temperature bulb is a device that varies its resistance with a change in temperature so that the balancing resistor dial can be calibrated to read out a process temperature.

When the bridge circuit is balanced (i.e., no current through galvanometer $G$), the circuit can be analyzed as two series resistance strings in parallel. The equal voltage ratios in the two branches of the Wheatstone bridge can be stated as

$$\frac{I_2 R_x}{I_2 R_s} = \frac{I_1 R_1}{I_t R_2}$$

Note that $I_1$ and $I_2$ can be canceled out in the previous equation, so that we can invert $R_s$ to the right side of the equation to find $R_x$ as follows:

$$R_x = R_s \frac{R_1}{R_2} \tag{4-11}$$

To illustrate the use of a Wheatstone bridge, let's consider the following example.

**EXAMPLE 4-8**

**Problem:** Assume a Wheatstone bridge with $R_1 = 1\ k\Omega$ and $R_2 = 10\ k\Omega$. If $R_s$ is adjusted to read 50 $\Omega$ when the bridge is balanced (i.e., galvanometer reading at zero), calculate the value of $R_x$.

**Solution:** We can use Equation 4-11 to determine $R_x$:

$$R_x = R_s \frac{R_1}{R_2}$$

$$R_x = 50\ \Omega\ \frac{1\ k\Omega}{10\ k\Omega} = 5\ \Omega$$

## Instrumentation Current Loop

The brief discussion on series resistive circuits was leading to a discussion on instrumentation current loops. The dc current loop shown in Figure 4-6 is used extensively in the instrumentation field to transmit process variables to indicators and/or controllers. It is also used to send control signals to field devices to manipulate process variables such as temperature, level, and flow. The normal current range used is 4 to 20 mA, and this value is converted to 1 to 5 V dc by a 250 $\Omega$ resistor at the input to controllers and indicators that are normally high input impedance ($Z_{in} > 10\ M\Omega$) electronic amplifiers that draw virtually no current.

There are two main advantages to using the 4 to 20 mA current loop. First, only two wires are required for each remotely mounted field transmitter, so a cost savings is realized on both labor and wire in installing field devices. The second advantage is that the current loop is not affected by electrical noise.

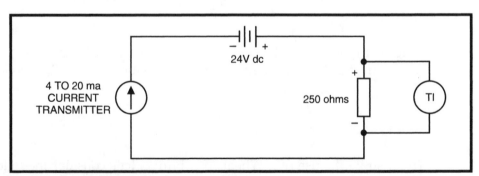

**Figure 4-6. Typical 4 to 20 MA current loop.**

## Selection of Wire Size

Several factors must be considered in selecting the wire size in a programmable controller application. One factor is the permissible power loss ($P = I^2R$) in the electrical line. This power loss is electrical energy being converted into heat. If the heat produced is excessive, damage to the conductors or system components might result. The use of large diameter (low wire gauge) conductors will reduce the circuit resistance and, therefore, the power loss. However, larger diameter conductors are more expensive than smaller ones, and they are more costly to install, so design calculations need to be made to select the proper wire size.

A second factor is the resistance of the wires in the circuit. For example, let's assume we are sending a full range 20 mA dc signal from a field instrument to a programmable controller input module located 2000 feet away as shown in Figure 4-7.

Assuming we are using 20 AWG wire, we can easily calculate the wire resistance ($R_w$). Using data from Table 4-2, we see that 20 AWG wire as a resistance of 10.4 $\Omega$ per 1000 ft at 25°C. So the resistance for the 4000 feet of wire is calculated as follows:

$$R_w = (10.4 \ \Omega / 1000 \ \text{ft})(4000 \ \text{ft}) = 41.6 \ \Omega$$

This resistance increases the load on the 24 Vdc power supply used to drive the current loop.

The voltage drop $V_w$ in the wire when the field instrument is sending 20 ma of current to the programmable controller analog input module can be calculated using Ohm's Law as follows:

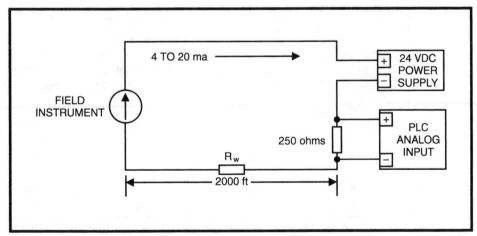

**Figure 4-7. Voltage drop in instrument loop.**

$$V_w = IR_w = (20 \text{ mA})( 41.6 \text{ } \Omega) = 0.832 \text{ volts}$$

A third factor is the current-carrying ability of the conductor. When current flows in a wire, heat is generated. The temperature of the wire will rise until the heat radiated away, or otherwise dissipated, is equal to the heat generated in the conductor. If the conductor is insulated, the heat produced is not so readily removed as it would be if the conductor were not insulated. So, to protect the insulation from excessive heat, the current flowing in the wire must be kept below a certain value. Rubber insulation will start to deteriorate at relatively low temperatures. Teflon and certain plastic insulation retain their insulating properties at higher temperatures, and insulations such as asbestos are effective at still higher temperatures.

Electrical cables might be installed in locations where the ambient temperature is relatively high. In these cases, the heat produced by external sources adds to the total heating of the electrical conductor. Therefore, allowances must be made in the design of a programmable controller system for the ambient heat sources encountered in industrial environments. The maximum allowable operating temperature of insulated wires and cables is specified in manufacturer's electrical design tables.

An example of the maximum allowable current-carrying capacities of copper conductor with three different types of insulation is given in Table 4-3. The current ratings listed in the table are those permitted by the National Electrical Code.

This table can be used to determine the safe and proper wire size in an electrical wiring application. An example problem will help to illustrate a typical wire sizing calculation.

| Conductor Size (AWG) | 60°C (140°F) Types: TW, UF | 75°C (167°F) Types: FEPW, RH, RHW, THHW, THW, THWN, XHHW, USE, ZW | 85°C (185°F) Type: V18 |
|---|---|---|---|
| 18 | . . . | . . . | . . . |
| 16 | . . . | . . . | 18 |
| 14 | 20 | 20 | 25 |
| 12 | 25 | 25 | 30 |
| 10 | 30 | 35 | 40 |
| 8 | 40 | 50 | 55 |
| 6 | 55 | 65 | 70 |

Table 4-3. Ampacities of Insulated Copper Conductor National Electrical Code (1).

**EXAMPLE 4-9**

**Problem:** Calculate the proper wire size for the 120 V AC feed to the programmable controller system shown in Figure 4-8, assuming a maximum ambient temperature of 60°C.

**Solution:** The total AC feed current $I_t$ is the sum of all the branch currents, or

$$I_t = I_1 + I_2 + I_3 + I_4$$
$$= 2\,A + 8\,A + 5\,A + 5\,A$$
$$= 20\,A$$

Using Table 4-3, we can see that the main power conductors (hot, neutral, and ground) must be 12 AWG.

## Power Supplies

A basic understanding of the design and operation of dc power supplies is important in programmable controller applications because of their wide use in control systems. DC power supplies use either

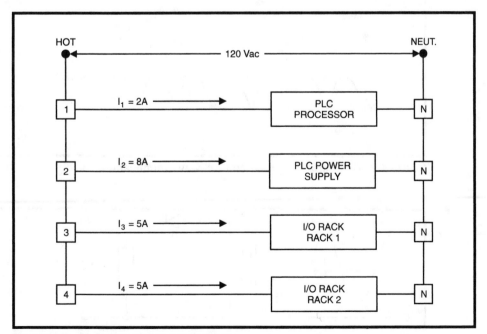

**Figure 4-8. PLC wire sizing application.**

half-wave or full-wave rectification depending on the power supply application. Figure 4-9 shows the output waveforms produced by half-wave and full-wave rectification. A schematic diagram of a half-wave rectifier is shown in Figure 4-10. In this circuit, the primary input voltage is 120 volts rms, and the secondary is given as 25 volts rms. The positive and negative cycles of the ac voltage across the secondary winding are in phase with the signal at the primary.

If we assume that the top of the transformer secondary is positive and the bottom is negative during a positive cycle of the input, then the diode is forward-biased and current flow is permitted through the load resistor, $R_L$, as indicated.

During the negative input cycle, the top of the transformer will be negative and the bottom will be positive. This voltage polarity reverse

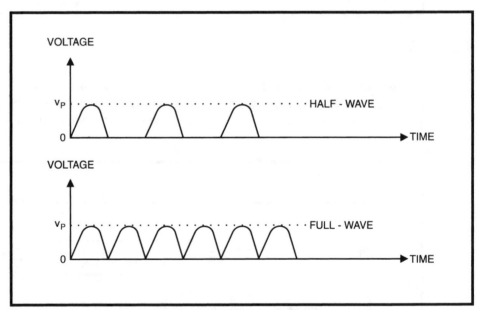

Figure 4-9. Comparison of half- and full-wave rectification.

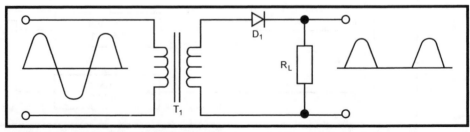

Figure 4-10. Half-wave rectifier circuit.

biases the diode. A reverse-biased diode represents an extremely high resistance and serves as an open circuit. Therefore, with no current flowing through $R_L$, the output voltage will be zero for this cycle.

To calculate the dc voltage of the half-wave rectifier, we first determine the peak voltage of one cycle. This is obtained by multiplying the rms input voltage by 1.414. The average value is then determined by multiplying the peak value by 0.637. Since only one cycle of the two input cycles appears in the output, this value must be divided by 2. Combining these values we find that 1.414 times 0.367 divided by 2 equals 0.45, or 45% of the rms ac value minus the diode voltage drop $V_d$ of 0.7 volt. A dc voltmeter would read this value across the load resistor, $R_L$ in the simple half-wave rectifier shown in Figure 4-10.

$$\text{Peak voltage } (V_p) = 1.414 \times V_{rms}$$

$$V_p = 1.414 \times 25\ V = 35.35\ V$$

$$\text{Average voltage } V_{av} = 0.367 \times V_p$$

$$V_{av} = 0.367 \times 35.35\ V$$

$$\text{DC output } (V_{dc}) = (V_{av}/2) - V_d$$

$$V_{dc} = (22.5\ V/2) - V_d$$

$$V_{dc} = (11.25 - 0.7)V = 10.55\ V$$

The basic full-wave rectifier using two diodes is shown in Figure 4-11. The cathodes of both diodes are connected to obtain a positive output. The anodes of each diode are connected to opposite ends of the transformer secondary winding. The load resistor ($R_L$) and the transformer center is connected to ground to complete the circuit.

In a full-wave rectifier, the current through $R_L$ is in the same direction for each alteration of the input, so that we obtain dc output for both halves of the sine-wave input, or full-wave rectification.

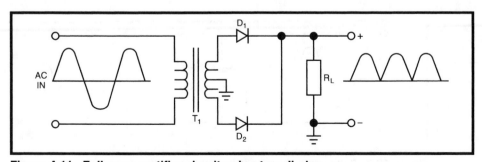

**Figure 4-11. Full-wave rectifier circuit using two diodes.**

The dc output voltage of a full-wave rectifier is 90% of the ac rms voltage appearing between the center tap and the other ends of the transformer. This voltage is determined by calculating the peak value $(V_p)$ of the rms voltage, then multiplying it by the average value $(V_{av})$. Since 1.414 times 0.637 equals 0.900, the potential dc output is 90% of the rms input. The dc output voltage appearing across $R_L$ of the circuit will be slightly less than 90% of the rms value because each diode has a voltage drop $(V_d)$ of 0.7 volt. This means that a dc voltmeter would read the rms value times 90% minus 0.7 volt across $R_L$.

The ripple frequency of the full-wave rectifier is also different from that of a half-wave circuit. Since each cycle of the input produces an output across $R_L$, the ripple frequency will be twice the input frequency. The higher ripple frequency and characteristic output of the full-wave rectifier are easier to filter than a similar half-wave output.

A bridge structure of four diodes is commonly used in power supplies to achieve full-wave rectification. In the rectifier shown in Figure 4-12, two diodes will conduct during the positive alteration and two will conduct during the negative alteration. A bridge rectifier does not require a center-tapped transformer as used in a two-diode, full-wave rectifier.

The dc output appearing across $R_L$ of the bridge circuit has a ripple frequency of 120 Hz and will have a dc voltage of a little less than 90% of the secondary rms value. Since each diode produces a 0.7-volt drop, the two diodes that conduct will reduce the output by two times 0.7, or 1.4 volts. The resulting dc output will be 90% of the rms value less 1.4 volts, so that in the circuit shown in Figure 4-12, the dc output voltage $(V_o)$ will be

$$V_o = 0.9 \times 25V - 1.4\ V = 21.1\ Vdc$$

Bridge rectifiers are commonly used in electronic power supplies for instruments because of their simple operation and desirable output. The diode bridge is generally housed in a single enclosure that has two

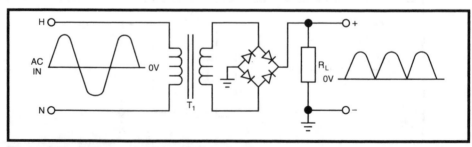

**Figure 4-12. Full-wave bridge rectifier circuit.**

input and two output connections. However, the output from a bridge rectifier still needs to be filtered to produce the smooth dc signal needed by most instrumentation circuits and devices.

The most common types of electronic devices and circuits encountered in programmable controller applications have been discussed. We now need to discuss some common electrical control devices.

# Electrical Control Devices

A wide variety of electrically operated devices are found in control applications. We will cover the most common devices in this section and give the electrical symbol used in drawings to represent each device.

## Electrical Relays

The most common control device encountered in control applications is the electromechanical relay. It is called electromechanical because it consists of electrically operated solenoid and mechanical contacts. The solenoid is normally a long thin wire wound into a close-packed helix around an iron armature; when an electric current is applied, a strong magnetic field is produced. The resulting magnetic force moves the armature that is connected to a common contact or set of common contacts when electrical power is applied. The contacts in the relay are used to make or break electrical connections in control circuits. This set of contacts is used to switch a current that can be much higher than the original control signal.

A typical relay is shown in Figure 4-13, with two sets of contacts. Each set consist of a common (COM) contact, a normally open (NO) contact, and a normally closed (NC) contact. Normally open and normally closed refer of the status of contacts when no electrical power is applied to the solenoid or when the relay is in the shelf position. For example, if there is no electrical continuity between a common contact and another contact in the set when the relay is on a storage shelf, this set of contacts is termed the normally open set of contacts.

The electrical schematic symbol for a NO set of contacts is given in Figure 4-14a and the symbol for the NC set is shown in Figure 4-14b. The standard symbol for the relay solenoid is a circle with two connection dots as shown in Figure 4-14c. This symbol will normally also include a combination of letter(s) and number(s) to identify each control relay on a control drawing. Typical designations are CR1, CR2, . . . CRn for the control relays on a schematic.

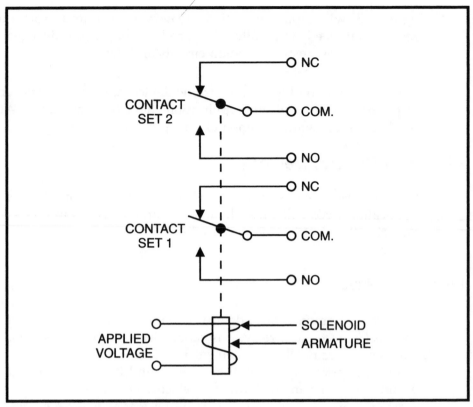

Figure 4-13. Typical electric relay.

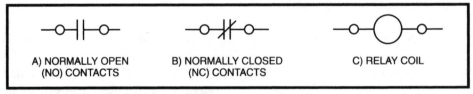

Figure 4-14. Electrical relay symbols.

## Electric Solenoid Valves

Another common electrically operated device encountered in process control is the solenoid valve. The solenoid valve is a combination of two basic functional units: an electromagnetic solenoid with its core and a valve body containing one or more orifices. Flow through an orifice is allowed or prevented by the action of the core when the solenoid is energized or deenergized.

Solenoid valves normally have a solenoid mounted directly on the valve body. The core is enclosed and free to move in a sealed tube called a core tube, thus providing a compact assembly. The electrical schematic symbol for a solenoid valve is given in Figure 4-15.

The three common types of solenoid valves are direct acting, internal pilot-operated, and manual reset. In direct-acting valves, the solenoid core directly opens or closes the orifice, depending upon whether the solenoid is energized or deenergized. The valve will operate from 0 psi inlet pressure to its maximum rated inlet pressure. The force required to open a solenoid valve is proportional to the orifice size and the pressure drop across the valve. As the orifice size increases, the force needed to operate the valve increases. To keep the solenoid coil small and to open large orifices at the same time, internal pilot-operated solenoid valves are used.

Internal pilot-operated solenoid valves have a pilot and bleed orifice and use the line pressure for operation. When the solenoid is energized, the core opens the outlet side of the valve. The imbalance of pressure causes the line pressure to lift the piston or diaphragm off the main valve orifice, opening the valve. When the solenoid is deenergized, the pilot orifice is closed and the full line pressure is applied to the top of the piston or diaphragm through the bleed orifice to close the valve. In some applications, the bleed orifice is replaced by a small manual valve to allow for controlling the speed required to open and close the valve. The manual rest type of solenoid valve must be manually positioned (latched). It will return to its original position when the solenoid is energized or deenergized, depending on the valve design. Three configurations of solenoid valves are normally available for process control applications: two, three, and four way. Figure 4-16 shows the

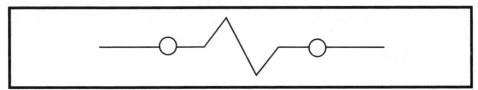

**Figure 4-15. Schematic symbol for solenoid valve.**

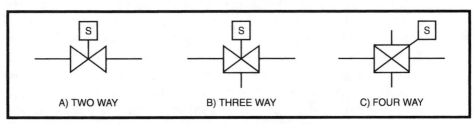

A) TWO WAY          B) THREE WAY          C) FOUR WAY

**Figure 4-16. Proces diagram symbols for solenoid valves.**

process diagram symbols used to represent two-, three-, and four-way solenoid valves. Two-way valves have one inlet and one outlet process connection. They are available in either normally open or normally closed configuration. Three-way solenoid valves have three process piping connections and two orifices (one orifice is always open and the other is always closed). These valves are normally used in process control to alternately apply air pressure to and exhaust air from diaphragm-operated control valves or single-acting cylinders. Four-way solenoid valves have four pipe connections: one for supply air pressure, two for process control, and one for air exhaust. These valves are normally used to control double acting actuators on control valves.

## Electrical Contactors

Electrical contactors are relays that are able to switch high current loads normally greater than 10 amps. They are used to repeatedly make and break electrical power circuits. An electrically operated solenoid is the most common operating mechanism for contractors. The solenoid is powered by low ampere AC voltage signals from control circuits. Instead of opening or closing a valve, the linear action of the solenoid coil is used to open or close sets of contacts with high power ratings. The higher rated sets of electrical contacts are used in turn to control loads such as motors, pumps, and heaters. The schematic symbol for a typical contactor is shown in Figure 4-17.

## Other Electrical Devices

Hand switches, instrument switches, control switches, push buttons, and indicating lights are some of the common devices encountered in process control. Switches and push buttons are used to control other devices, such as solenoid valves, electric motors, motor starters, heaters, and other process equipment.

Hand switches come in a variety of designs and styles and they can be mounted on control panels or in the field near the process equipment. Instrumentation switches are generally field mounted and are used to sense pressure, temperature, flow, liquid level, and position.
To make the reading of control drawings easier, a set of standardized symbols was developed by the American National Standards Institute (ANSI). The most common symbols used in control diagrams are shown in Figure 4-18; they are based on ANSI Y32.2 for electrical controls.

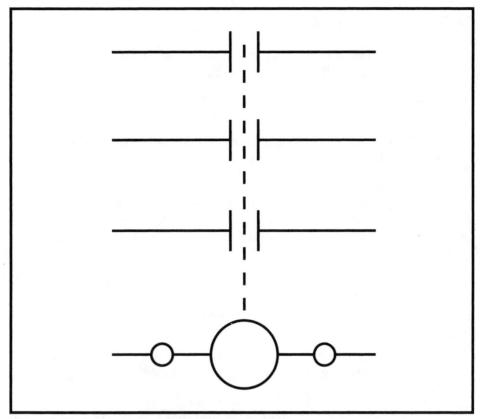

**Figure 4-17. Common electrical contactor symbol.**

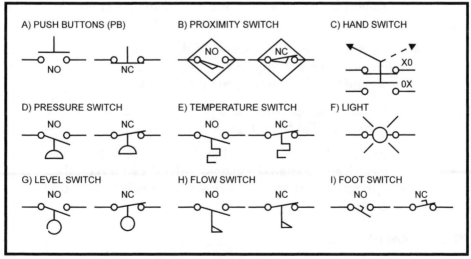

**Figure 4-18. Common electrical and instrument symbols.**

## EXERCISES

4.1   A charge of 15 coulombs moves past a given point every second. What is the current flow?

4.2   Calculate the current in a conductor if $30 \times 10^{18}$ electrons pass a given point in the wire every second.

4.3   Calculate the resistance of 500 feet of silver wire with a diameter of 0.001 inch.

4.4   The resistance to ground on a faulty underground telephone line is 20 $\Omega$. Calculate the distance to the point where the wire is shorted to ground, if the line is a 22 AWG copper conductor and the ambient temperature is 25°C.

4.5   Find the total current ($I_t$) in a series circuit with two 250 $\Omega$ resistors if the applied voltage is 24 Vdc. Also find the voltage drop, $V_1$, across resistor $R_1$ and the voltage drop, $V_2$, across $R_2$.

4.6   In the parallel resistance circuit of Figure 4-4, assume that $R_1$ = 100 $\Omega$, $R_2$ = 200 $\Omega$, and $V_t$ = 100 Vdc. Find $I_1$, $I_2$, and $I_t$.

4.7   Assume that the Wheatstone bridge circuit shown in Figure 4-5 is balanced (i.e., $I_g$ = 0) and that $R_1$ = 2 k$\Omega$, $R_2$ = 10 k$\Omega$, $R_s$ = 2 k$\Omega$ and $V_t$ = 12 Vdc. Calculate the value of $I_1$, $I_2$, and $R_x$. Also, what are the voltage drops across $R_1$ and $R_2$?

4.8   Find the dc output voltage for the half-wave rectifier circuit shown in Figure 4-10, if the peak voltage across the secondary of the transformer $T_1$ is 40 V.

4.9   Find the dc output voltage (V dc) for the full-wave rectifier circuit shown in Figure 4-11, if the peak voltage across the secondary of the transformer $T_1$ is 20 V.

4.10  Explain the operation and purpose of electromechanical relays and list three typical applications.

4.11  Explain the operation of the three common types of electrically operated solenoid valves. List a typical application for each type of solenoid valve in process control.

## BIBLIOGRAPHY

1.   Malvino, A. P., *Electronic Principles*, Second Edition, McGraw-Hill Book Co., 1979.
2.   Grob, B., *Basic Electronics*, Fifth Edition, McGraw-Hill Book Co., 1984.

3. Budak, A., *Passive and Active Network Analysis and Synthesis*, Houghton Mufflin Co., 1974.

4. Motorola Semiconductor Products Inc., *Zener Diode Handbook*, May 1967.

5. Bogart, T. F. Jr., *Electric Circuits*, Macmillan Publishing Co., 1988.

6. Boylestad, R., Nashelsky, L., *Electric Devices and Circuit Theory*, Third Edition, Prentice—Hall, Inc., 1982.

7. Bartkowiak, R. A., *Electric Circuit Analysis*, Harper & Row, Publishers, Inc., 1985.

8. Hughes, T. A., *Measurement and Control Basics*, 2nd Edition, Instrument Society of America, 1995.

9. *The National Electrical Code Handbook*, Sixth Edition National Fire Protection Association, 1996.

# 5

# Input/Output Systems

## Introduction

The input/output (I/O) system provides the physical connection between the process equipment and the central processing unit (CPU) or simply "processor." It uses various interface circuits and/or modules to sense and measure physical quantities of the process, such as motion, level, temperature, pressure, current, and voltage. Based on the status sensed or values measured, the control program in the processor activates various devices, such as valves, motors, pumps, and alarms, to exercise control over a machine or process using output modules.

These I/O circuits or modules are mounted in equipment housings or, in the case of micro-PLCs, in part of the programmable logic controller (PLC) housing. In most PLC housings, any I/O module can be inserted into any I/O slot, and the housings are designed so that the I/O modules can be removed without turning off the ac power or removing the field wiring. Most I/O modules use printed circuit board technology, and the circuit boards have an edge connector that can be inserted into a plug in the backplane connector of the rack. This backplane is a printed circuit card that contains the parallel communication lines or bus to the processor and the dc voltages that are required to power the logic and interface circuits in the I/O modules.

## Discrete Inputs

Discrete inputs are the most common class of field signals in a PLC system. This type of interface connects a field input device, which will provide an input signal that is separated and distinct in nature. This characteristic limits the discrete input interfaces to sensing signals that are on/off or open/closed (or equivalent), to a switch closure. To the interface circuit or module, all discrete inputs are essentially devices with two states. Table 5-1 lists the most common discrete input devices encountered in process control applications.

When in operation, if an input switch is closed, the input circuit senses the supplied voltage and converts it to a logic-level signal acceptable to the CPU to indicate the status of the device. A logic 1 indicates ON or CLOSED, and a logic 0 indicates OFF or OPENED.

## Discrete Outputs

Discrete output control is limited to devices that require switching only one of two states, such as on/off, open/closed, or extended/retracted. Table 5-2 lists the most common discrete output field devices encountered in process and machine control applications.

In operation, the output interface circuit switches the supplied control voltage that will energize or deenergize the device. If an output is turned ON through the control program, the supplied control voltage is switched by the interface circuit to activate the referenced (addressed) output device.

## I/O Signal Types

Each discrete input and output signal is powered by some field or panel supplied voltage source (e.g., +5 Vdc, 120 Vac, 24 Vdc, etc.). I/O

| | |
|---|---|
| Selector switches | Motor starter contacts |
| Thumbwheel switches | Limit switches |
| Temperature switches | Pressure switches |
| Flow switches | Hand switches |
| Level switches | Proximity switches |
| Valve position switches | Relay contacts |
| Starter auxiliary contacts | Limit switches |
| Pushbuttons | Photoelectric senors |

Table 5-1.  Field Input Devices.

| | |
|---|---|
| Annunciators | Electric valves |
| Alarm lights | Alarm horns |
| Electric control relays | Solenoid valves |
| Electric fans | Motor starters |
| Indicating lights | Heater starters |

Table 5-2.  Discrete Output Field Devices.

interface circuits are available at various ac and dc voltage ratings, as listed in Table 5-3.

The most common voltage level used in industrial applications is 120 Vac because this voltage is readily available in industrial plants. However, +24 Vdc is also widely used for safety reasons because there is less risk of injury with +24 Vdc than with 120 Vac.

## Sinking and Sourcing Operations

Sink and source operations refer to the electrical configuration of an electronic circuit in a device, whether it is an input module or a field input device. If the device provides current during its ON state, the device is said to be sourcing current. Conversely, if the device receives current in the ON or true state, it is said to be sinking current. Therefore, we can have sinking and sourcing field devices as well as sinking and sourcing input modules. However, the most common configuration using PLC applications is the sourcing field input device and the sinking input module.

Potential interface problems can arise if the user does not design the I/O system to properly match sinking and sourcing operations for the devices in the system. The user must use sinking input circuits in an input module if the field devices connected to the module are sourcing devices. Conversing, sourcing input circuits must be used if they are connected to a sinking field device for proper operation. An interface problem would arise, for instance, if an input module is designed for sink operation and all input devices except one are operating in a source configuration. The sinking input device may be ON, but the module would not detect the ON signal, even though a voltage could be measured across the module's input terminals. There is also the potential for damage to the mismatched field device and input circuit in the I/O module.

## Discrete AC Voltage Input Circuits

A block diagram of a typical alternating current (ac) voltage discrete input circuit is shown in Figure 5-1. Discrete input circuits vary widely

| | |
|---|---|
| + 5 volt dc | 120 volt ac or dc |
| +12 volt dc | 230 volt ac or dc |
| +24 volt ac or dc | 100 volts dc |
| +48 volt ac or dc | Dry contacts |

**Table 5-3. Typical Discrete I/O Signal Types.**

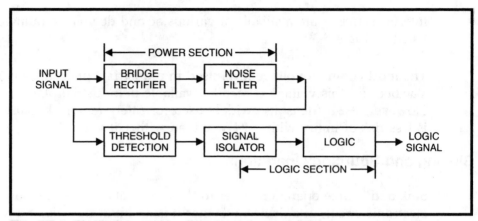

**Figure 5-1. Block diagram of a discrete ac input circuit.**

among PLC manufacturers, but, in general, all discrete input circuits operate in a manner similar to that described in Figure 5-1.

The input voltage circuit is composed of two primary parts: the power section and the logic section. The power and logic sections of the circuit are normally coupled with a circuit that electrically isolates the input power section from the logic circuits. This electrical isolation is very important in a normally noisy industrial environment. The main problem with the early application of computers to process control was that the inputs and outputs were not designed for the harsh industrial environment.

Figure 5-2 shows a typical discrete ac input circuit. The power section of the circuit performs the function of converting the incoming voltage (115 Vac, 230 Vac, etc.) from a field input device, such as described in Table 5-1, to a logic-level signal to be used by the PLC processor during its control program control. The bridge rectifier circuit converts the ac incoming signal to a dc level that is sent to a filter circuit, which protects against electrical noise on the input power line. This RC filter circuit causes a signal delay that is typically 10–25 msec. The threshold circuit uses a zener diode ($Z_d$) to detect whether the incoming signal has reached the proper voltage level for the specified input rating. If

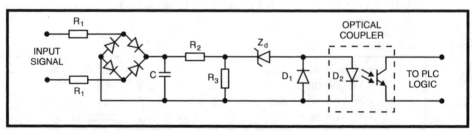

**Figure 5-2. Typical ac voltage discrete input circuit.**

the input signal exceeds and remains above the threshold voltage for a duration of at least the filter delay, the signal will be accepted as a valid input.

When a valid signal has been detected, it is passed through the isolation circuit, which completes the electrically isolated transition from ac or dc voltage to a logic level voltage. The logic level signal from the isolator is used by the logic circuit and made available to the processor via the data communications bus. Electrical isolation is provided so that there is no electrical connection between the field device (power) and the controller (logic). This electrical separation helps prevent large voltage spikes from damaging the logic side of the interface (or the controller). The coupling between the power and logic sections is normally provided by an optical coupler as shown in Figure 5-2 or a pulse transformer. Electrical isolation is one of the reasons the programmable controller has gained very wide acceptance in the process industries.

In small, medium, and large PLC systems, these discrete input circuits are mounted together on a single circuit board and installed in an input module. Each input module will normally have 4, 8, 16, or 32 input circuits in one unit.

## Discrete AC Input Modules

Most discrete ac input modules will have a signal indicator to signify that the proper input voltage level is present (a switch is closed). A light-emitting diode (LED) indicator is normally used to indicate the status of the input. These indicating lights are an important aid during system start-up and troubleshooting. A discrete ac input connection diagram is shown in Figure 5-3.

The letters "ACI-120" are used to designate a 120 Vac discrete input module. This designation will be used in this book on equipment lists, wiring diagrams and system drawings. We will also use "ACI-220" to designate a 220 Vac input module. All classes of discrete input module will use the letters "DI" on drawings.

## Direct Current (dc) Input Modules

The dc voltage input modules convert discrete on/off direct current inputs to logic level signals compatible with the programmable controller. They are generally available in three voltage ranges: 12, 24, and 48 volts dc. Typical instruments that are compatible with the module include limit switches, valve position switches, push buttons, dc proximity switches, float switches, and photoelectric sensors.

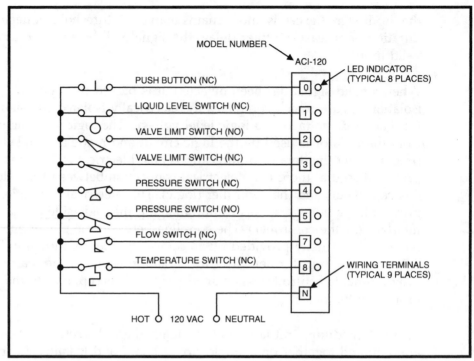

**Figure 5-3. Typical discrete ac input module wiring diagram.**

The model number for the 24 Vdc discrete input module is DCI-24 in this presentation. We will also designate a 12-Vdc input module as "DCI-12" and a 48-Vdc input module as "DCI-48."

The wiring diagram for a dc input module would be the same as the wiring of the ac input module shown in Figure 5-3 except the supply voltage would be a direct current voltage instead of an ac voltage. The ac hot voltage signal sent to the field devices would be replaced by a positive dc voltage and the neutral terminal on the module would be replaced by a common or negative dc voltage terminal.

## Transistor-Transistor Logic (TTL) Input Modules

The TTL input modules allow the controller to accept signals from TTL-compatible devices, including solid-state controls and sensing instruments. TTL inputs are also used for interfacing with some 5-V dc level control devices and several types of photoelectric sensors. The TTL interface has a configuration similar to the dc input modules; however, the input delay time caused by filtering is generally much shorter. TTL input modules normally require an external +5 Vdc power supply.

## Isolated Input Module

Input and output modules usually have a common return line connection for each group of inputs or outputs on a single module.

However, sometimes it may be required to connect an input device of different ground levels to the controller. In this case, isolated input modules with separate return lines for each input circuit are available (ac or dc) to accept these signals. The operation of the isolated interface is the same as the standard discrete I/O, except the common of each input is separated from the other commons in the module. The result is that the isolated input module requires twice as many input connection terminals, so it can accommodate only half the inputs in the same physical space as shown in Figure 5-4. The module number for the isolated 120 Vac input module is selected as "IACI-120" and the isolated 220 Vac input module can be designated as "IACI-220."

## Discrete AC Output Circuit

Figure 5-5 shows a block diagram of a typical discrete ac output circuit. AC output circuits vary widely among PLC manufacturers. The block diagram, however, describes basic operation of most ac output circuits. The circuit consists primarily of the logic and power sections, coupled by an isolation circuit. The output interface can be thought of as a simple switch through which power can be provided to control the output device.

During normal operation, the processor sends to the logic circuit the output status determined by the logic program. If the output is energized, the signal from the processor is fed to the logic section and passed through the isolation circuit, which will switch the power to the field device.

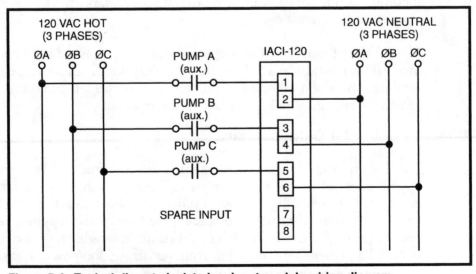

**Figure 5-4. Typical discrete isolated ac input module wiring diagram.**

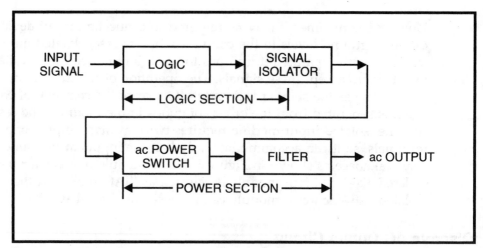

**Figure 5-5. Block diagram of a discrete ac output circuit.**

The switching section generally uses a Triac™ or a silicon-controlled rectifier (SCR) to switch the power. The ac switch is normally protected by an RC snubber and often a metal oxide varistor (MOV), which is used to limit the peak voltage to some value below the maximum rating and also to prevent electrical noise from affecting the module operation. A fuse may be provided in the output circuit to prevent excessive current from damaging the ac switch. If the fuse is not provided on each circuit in the module, it should be added to the exterior of each output circuit.

## Discrete AC Output Module

In small, medium, and large PLC systems these discrete output circuits are mounted together on a single circuit board and installed in an input module. Each input module will normally have 4, 8, 16, or 32 input circuits on the circuit card.

As with input modules, the output module may provide light-emitting diode (LED) indicators to show the operating logic. An ac output module connection diagram is illustrated in Figure 5-6. Note that the switching voltage is field-supplied to the module.

## Direct Current (dc) Output Module

The dc output module is used to switch direct current loads. Functional operation of the dc output is similar to the ac output; however, the power circuit generally employs a power transistor to switch the load. Like Triacs, transistors are also susceptible to excessive applied voltages and large surge currents, which could result in overheating and a short circuit. To prevent this condition from occurring, the power transistor will normally be protected with a fuse.

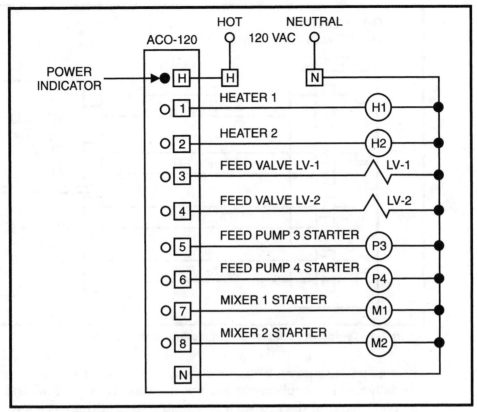

**Figure 5-6. Typical discrete output module wiring diagram.**

The wiring diagram for a direct current output module would be the same as the wiring of the ac output module shown in Figure 5-6 except the supply voltage would be a dc voltage instead of an ac voltage. The terminal of the ac hot line would be replaced by a positive dc voltage terminal and the ac neutral terminal would be replaced by a ground or negative dc voltage terminal on a dc input module.

## Dry Contact Output Module

The contact output module allows output devices to be turned ON or OFF by a normally open (NO) or normally closed (NC) set of relay contacts. The advantage of relay or dry contact outputs is that you have electrical isolation between the power output signal and the logic signal. Figure 5-7 shows a dry contact output module with four normally open contacts controlling the starting and stopping of two field-mounted variable speed motor drive units. It also shows that there is complete electrical isolation when the contacts are open. So there is no current flow when the output is off. The standard AC output module uses a triac in the output circuit that has a small leakage current that can cause problems in some applications. In these applications the Dry Contact Output module should be used.

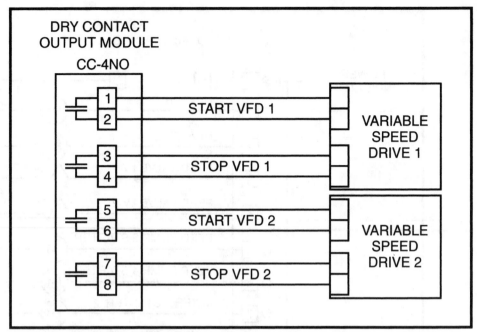

**Figure 5-7. Typical dry contact output module wiring diagram.**

The contact output can be used to switch either ac or dc loads but are normally used in applications such as isolating between PLC and complex electrical devices such as variable speed drives as shown in Figure 5-7. High current rated contacts (i.e., over 10 amps) are also available for applications that require switching of higher currents.

## TTL Output Module

The TTL output module allows the controller to drive output devices that are TTL compatible, such as seven-segment LED displays, integrated circuits, and various +5 Vdc logic-based devices. These modules generally require an external +5 Vdc power supply with specific current requirements.

## Isolated AC Output Module

An isolated ac output interface is shown in Figure 5-8. Note that the isolated AC output module is driving three different loads (A, B, C) that are connected to three different phases of ac power. The advantage of these modules is that we do not have to be concerned with the various ac voltage phases in our process plant. The disadvantages are that we increase the amount of wiring required and decrease the number of available inputs per module by a factor of two (2). In the application shown in Figure 5-8, three phases of 120 Vac power are used to turn on three separate motor starters for Pumps A, B, and C. This is a typical application for isolated ac output modules.

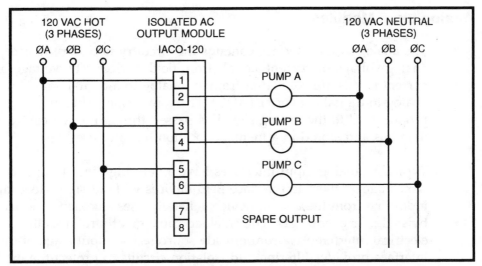

**Figure 5-8. Typical isolated ac output module wiring diagram.**

# Analog I/O Modules

The availability of low cost integrated circuits and industrial grade electronic circuits greatly increased the capabilities for analog circuits in programmable controller. This expanded capability led to the introduction of sophisticated analog input/output modules.

Analog input modules allow measured quantities to be recieved from process instruments and other devices that provide analog data, while analog output modules allow control of devices that require a continuous analog signal. The analog I/O will allow monitoring and control of analog voltages and currents, which are compatible with many sensors, motor drives, and process instruments. With the use of analog and special purpose I/O, most process variables can be measured or controlled with appropriate interfacing. Table 5-4 lists typical I/O devices that are interfaced to programmable controllers using analog modules.

| Analog Input Devices | Analog Output Devices |
|---|---|
| Flow transmitters | Electric motor drives |
| Pressure transmitters | Analog meters |
| Temperature transmitters | Chart data recorders |
| Analytical transmitters | Process controllers |
| Position transmitters | Current-to-pneumatic transducers |
| Potentiometers | Electrical-operated valve |
| Level transmitters | Variable speed drives |
| Speed instruments | |

**Table 5-4. Typical Analog I/O Field Devices.**

## Analog Input Modules

The analog input interface contains the circuitry necessary to accept analog voltage or current signals from field devices. The voltage or current inputs are converted from an analog to a digital value by an analog-to-digital converter (ADC). The conversion value, which is proportional to the analog signal, is passed through to the controller's data bus and stored in a memory location for later use.

Typically, analog input interfaces have a very high input impedance, which allows them to interface field devices without signal loading. The input line from the analog device generally uses shielded conductors or twisted pair conductors. The shielded cable greatly reduces the electrical interferences from outside sources. The input stage of the interface provides filtering and isolation circuits to protect the module from additional field noise. A typical analog input connection is illustrated in Figure 5-9. In the example shown, the analog input module is providing the DC voltage required by the field transmitters.

Most analog modules are designed to sense up to 16 single-ended or 8 differential analog input signals representing flow, pressure, level, etc. It then converts them to a proportional 12-bit binary or four-digit BCD value. Inputs to a particular module must, in general, be all single-ended or all differential, and the type of signal is either hardware or software selectable. The converted signals are stored in memory in the module and are sent to the processor memory in groups or blocks of data.

The control program uses configuration data to set up the analog module. Typical configuration information includes range selection (i.e., +1 to +5 Vdc, 4 to 20 mA, etc.) and signal scaling.

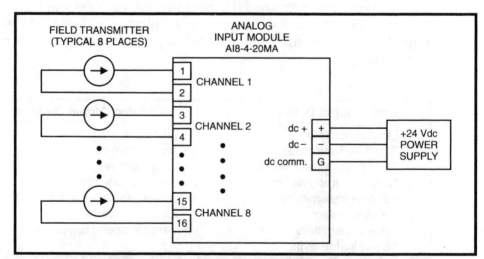

**Figure 5-9. Typical 8-channel analog input module wiring diagram.**

## Analog Output Modules

The analog output modules receive data from the central processing unit of the PLC, which is translated into a proportional voltage or current to control an analog field device. The digital data is passed through a digital-to-analog converter (DAC) and sent out in analog form. Isolation between the output circuit and the logic circuit is generally provided through optical couplers. These output modules normally require an external power supply with certain current and voltage requirements.

# Special Purpose Modules

There are a wide variety of special purpose I/O modules used in PLC systems. One PLC manufacturer has over 120 different types of I/O module. We will discuss some of the more common special purpose modules such as BCD I/O modules, Encoder/Counter Input Modules, and High Speed Pulse Input Modules.

### Binary Coded Decimal Input Modules

The binary coded decimal (BCD) input modules provide parallel communication between the processor and input devices, such as thumbwheel switches. This type of module is generally used to input parameters into specific data locations in memory to be used by the control program. Typical parameters are timer and counter presets, process control set point values, and batch number.

These modules generally accept voltages in the range of +5 V dc (TTL logic level) to 24 V dc and are grouped in a module containing 16 or 32 inputs which correspond to one or two data registers. Data manipulation instructions, such as GET data word or Block Transfer Read, are used to access the data from the register input interface. Figure 5-10 illustrates a typical device connection for a BCD input module. In this module, each bit (1 or 0) from the thumbwheel controls one bit location in a word location in the PLC.

### BCD Output Modules

This digital module provides parallel communication between the processor and an output device, such as a seven-segment LED display or a BCD display. The BCD output modules can also be used with TTL logic loads that have low current requirements. The BCD output module generally provides voltages that range from +5 Vdc (TTL) to +30 Vdc and have 16 or 32 output lines (one or two data words).

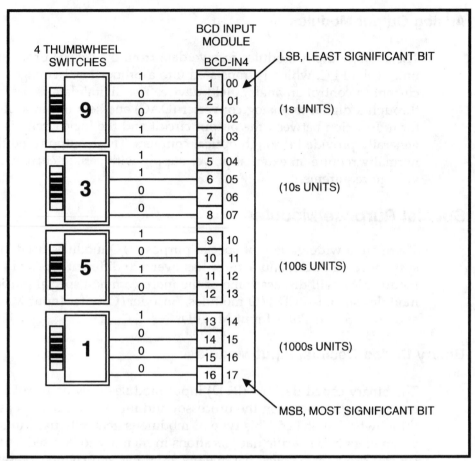

**Figure 5-10. Typical BCD input module wiring diagram.**

When information is sent from the processor through a data transfer instruction, the data is latched in the module and made available at the output terminals of the module.

## Encoder/Counter Input Module

The encoder/counter input module provides a high speed counter, external to the processor, which responds to input pulses sensed at the interface. This counter's operation is normally independent of either program scan or I/O scan. The reason for this is fairly simple: if the counter were dependent upon the PLC program, high speed pulses would be missed during a program scan. Typical applications of the encoder/counter interface are operations that require direct encoder input to a counter, which is capable of providing direct comparison outputs.

The encoder/counter input module accepts input pulses from an incremental encoder, which provides pulses that signify position when the encoder rotates. The pulses are counted and sent to the processor. Absolute encoders are generally used with interfaces that receive BCD or Gray code data, which represents the angular position of the mechanical shaft being measured.

During normal operations, the modules receive input pulses, which are counted and compared with a preset value selected by the operator. The counter input module normally has an output signal available, which is energized when the input and preset counts are equal; however, this is not needed in most PLCs. Since the data is available in the CPU, the programmer can use a comparison function to drive an output in the control program.

The data communication between the encoder/counter interface and the CPU is bidirectional. The module accepts the preset count value and other control data from the CPU and transmits data and status to the PLC memory. The output controls are enabled from the control program, which instructs the module to operate the outputs according to the count values received. The CPU, using the control program, enables and resets the counter operation.

## Pulse Counter Input Modules

The pulse counter input modules are used to interface with field instruments that generate pulses such as positive displacement (PD) flowmeters and turbine type flowmeters.

# Intelligent I/O Modules

In the previous sections, we discussed discrete, analog, and special purpose I/O modules that will normally cover 90% of the I/O applications encountered in PLC systems. However, to process certain types of signals or data efficiently, microprocessor-based intelligent modules are required. These intelligent interfaces include those that condition input signals, such as thermocouple modules, or other signals that cannot be interfaced using standard I/O modules. These intelligent modules can perform complete processing functions, independent of the CPU and the control program scan. In this section, we will discuss the most commonly available intelligent modules, such as thermocouple input, stepping motor output, and control loop modules.

## Thermocouple Input Modules

A thermocouple (T/C) input module is designed to accept inputs directly from a T/C and provides cold junction compensation to correct for changes in cold junction temperatures. The operation of this type of module is similar to the standard analog input with the exception that low level millivolt signals are accepted from the T/C (approximately 43 mV at maximum temperature for a J-type T/C). These signals are filtered, amplified, and digitized through an A/D converter and then normally sent to a built-in microprocessor to perform linearization of the millivolt input signal to convert the signal to a temperature value. Finally the temperature value is sent to the CPU on command from a program instruction. The temperature data is used by the PLC control program to perform temperature control and/or indication.

## Stepping Motor Module

The stepping motor module generates a pulse train that is compatible with stepping motor translators (see Figure 5-11). The pulses sent to the translator normally represent distance, speed, and direction commands to the motor.

The stepping motor interface accepts position commands from the control program. Position is determined by the preset count of output pulses; a forward or reverse direction command; and the acceleration or deceleration command for ramping control, which is determined by the rate of output pulses. These commands are generally specified during

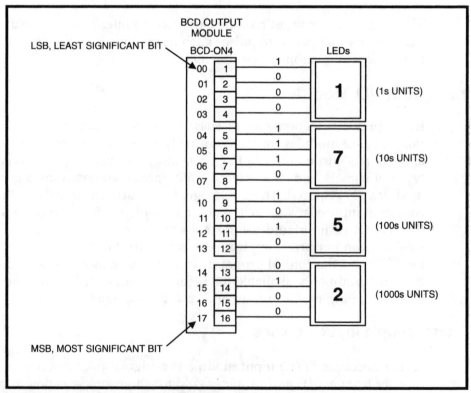

**Figure 5-11. Typical BCD output module wiring diagram.**

program control, and, once the output interface is initialized by a start, it will output the pulses according to the PLC program. Once the motion has started, the output module will generally not accept any commands from the CPU until the move is completed. Some modules may offer an override command that will reset the current position, which must be disabled to continue operation. The module also sends data regarding its status to the PLC processor. A typical dc stepping motor connection diagram is shown in Figure 5-12.

## Control Loop Module

The control loop module is used in closed-loop control where the proportional-integral-derivative (PID) control algorithm is required. Some manufacturers call this interface the PID module, and it is typically applied to any process operation that requires continuous closed-loop control. The control algorithm implemented by this module is represented by the following equation:

$$V_{out} = K_p e + K_i \int edt + K_d \frac{de}{dt} \tag{5-1}$$

where

$$
\begin{aligned}
V_{out} &= \text{output control variable,} \\
e &= PV - SP = \text{error,} \\
K_p &= \text{the proportional gain,} \\
K_i &= \text{the integral gain, and} \\
K_d &= \text{the derivative gain.}
\end{aligned}
$$

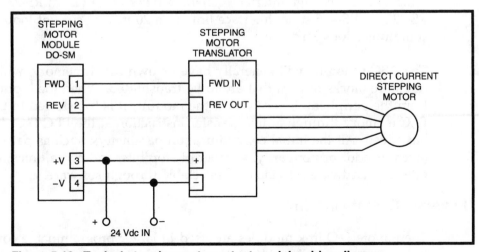

**Figure 5-12. Typical stepping motor output module wiring diagram.**

The PID module receives the process variable (PV), compares it to the set point (SP) selected by the operator, and computes the error difference. The operator or control engineer will determine the values of the control parameters ($K_p$, $K_d$, and $K_i$), based on the application.

The information sent to the PID module from the PLC processor is primarily the control parameters and set points. Depending on the PID module used, data can be sent to describe the update time, which is the period in which the output variable ($V_{out}$) is updated and the error dead band, which is a quantity that is compared to the error signal. If the error is less than or equal to the signal error, no update takes place. Some modules provide square root extraction of the process variable, which can be used to obtain a linearized scaled output for use on flow control loops.

# Communications Modules

The most common types of communication modules used in PLC systems to communicate among system components are the field bus interface module, the ASCII module, universal remote I/O link module, the serial communication module, PCMCIA interface card, the Ethernet interface module, and the fiber optic converter module.

## ASCII Communications Module

The ASCII communications module is used to send and receive alphanumeric data between peripheral equipment and the controller. Typical peripheral devices with ASCII I/O include printers, digital displays instruments, etc. This special I/O module, depending on the manufacturer, is available with a communications circuitry interface that includes onboard memory and a dedicated microprocessor. The information exchange interface generally takes place via an RS-232C, RS-422, or RS-485 serial interface link, or a 20 mA dc current loop communications link.

The ASCII module will generally have its own RAM memory, which can store blocks of data that are to be transmitted. When the input data from the peripheral is received at the module, it is transferred to the PLC memory through a data transfer instruction at the PLC I/O data bus speed. All the initial communication parameters, such as parity (even or odd) or nonparity, number of stop bits, and communication rate, are hardware selectable or selectable through software.

## Universal Remote I/O Link

The remote I/O link modules are used in larger programmable controller systems to allow I/O subsystems to be remotely located from the processor. The remote subsystem is used to interface with a unit process using a standard I/O rack and the required I/O modules. The rack will include a dc power supply to drive the internal circuitry of the I/O modules and a remote I/O adapter module that provides the communications with the processor unit. The I/O capacity of a single subsystem normally ranges from 32 to 256 points.

The subsystems are normally connected to the processor using a bus or star configuration. The distance from the processor to a given remote I/O rack normally ranges from 1000 feet to several miles, depending on the programmable controller type.

Remote I/O arrangements offer large cost savings on wiring materials (wire and conduit) and labor for large control systems in which the field instrumentation is in clusters at several remote process areas. If the processor is located in a main control room or some other central location, only the communication cable needs to be run between the processor and the field I/O racks, instead of hundreds or thousands of field wires.

Remote I/O arrangements also have the advantage of allowing subsystems to be installed and tested independently, as well as allowing maintenance and troubleshooting on individual stations while other units continue to operate.

## Serial Communications Module

The serial data communications module is normally used to communicate between the programmable controller and an intelligent instrument, with a serial output, such as a weigh scale with a serial communication port. This serial communication module generally has 2 to 4 serial ports to connect to RS-232, RS-422, and RS-485 standard communication interfaces.

## PCMCIA Interface Card

The Personal Computer Memory Card International Association (PCMCIA) developed a standard for a credit card-sized personal computer interface card. The PCMCIA standard defines an architecture and communications method for these PC interface cards. The interface cards developed under release 2.0 of the standard can be used for both data storage and I/O communication. PLC manufacturers developed PCMCIA cards to be installed in notebook personal computers that could communicate with the PLC processor or data highway to perform PLC software and troubleshooting functions.

These PCMCIA interface cards come with diagnostic software to verify proper operation of the card and connection to the PLC communication network.

## Ethernet Communications Modules

Ethernet interface modules are designed to allow a number of programmable controllers and other computer-based devices to communicate over a high speed plant ethernet communication network. This plant local area network (LAN) is able to transfer data and control information from one system to another at a high data transmission rate. Therefore, the control of an industrial facility can be distributed over a large number of programmable controllers, computers, and intelligent devices. In such a system, information is easily exchanged between control systems, but each system can independently control its part of the industrial plant. This greatly improves the reliability of the plant control system, since sections of the plant can be down for modification or maintenance, but the remaining parts of the plant can continue to operate and produce product.

## Fiber Optic Converter

The fiber optic converters transform electrical signals to light signals and transmit these signals through fiber optic cables. At the other end of the cable, a second fiber optic converter transforms light signals to electrical signals for use by the PLC system.

# Designing I/O Systems

To correctly design I/O systems, the programmable controller manufacturer's specifications must be consulted and followed to prevent faulty operation or equipment damage. These specifications place limitations not only on the module but also on the field equipment it operates. The specifications fall into three categories: electrical, mechanical, and environmental.

## Electrical Specifications

The typical electrical specifications for I/O modules include the following: (1) input voltage rating, (2) input current rating, (3) input threshold voltage, (4) output voltage rating, (5) output current rating, (6) output power rating, and (7) backplane current requirements.

The *input voltage (ac or dc) rating* specification lists the magnitude and type of signal the module will accept. In some cases, a range of input voltages instead of a fixed value is stated in the specification. In this case, the maximum and minimum acceptable working voltages for continuous operation are listed. For example, the working voltage for a 120 Vac input module might be listed as 95 to 135 Vac.

The *input current rating* defines the minimum input current required at the module's rated voltage that the field device must be capable of supplying to operate the input module circuit.

The *input threshold voltage is* the voltage at which the input signal is recognized as being ON or true. Some input modules also have an OFF voltage value at which the input is OFF or false. For example, the ON voltage for TTL input modules is defined as 2.8 Vdc or greater and the OFF voltage is defined as any voltage less than 0.8 Vdc.

The *output voltage rating* specifies the magnitude and type of voltage source that can be controlled within a stated tolerance. An output module rated at + 24 Vdc, for example, might have a working range of 20 to 28 Vdc.

The *output current rating* defines the maximum current that a single output circuit in a module can safely carry under load. This current rating is normally specified as a function of the output circuit component's electrical and heat dissipation characteristics at an ambient temperature range (typically 0° to 60°C). As the ambient temperature increases, the output current is normally derated. Exceeding the output current rating can result in a permanent short circuit or other damage to the output module.

The *output power rating* specifies the maximum total power that an output module can dissipate with all outputs energized. The output power rating for a single output is calculated by multiplying the output voltage rating by the output current rating. For example, if a 120 Vac output module has a current rating of 2 amps, the power rating is $P = I \times V$ or $P = 2\,A \times 120\,V = 240$ watts.

The *backplane current requirement* lists the current demand that a particular I/O module internal circuitry places on the rack power supply. The system designer must add up the current requirements of all the installed modules in an I/O rack and compare the value with the maximum current that can be supplied to determine if the power supplied is correct. If the rack power supply current is too low, intermittent and faulty operation of the system will result.

Some typical specifications for I/O modules are given in Table 5-5.

An example problem will illustrate a typical I/O system design application.

**EXAMPLE 5-1**

**Problem:** Calculate the backplane current requirements for the program-mable controller system with the following modules: five dc input modules, model number DCI-24; three dc output modules, model number DCO-24; two analog input modules, model number AI-4-20MA; and one analog output module mounted, model number AO-4-20MA mounted in a PLC rack. Assume the rack power supply and the backplane power bus are both rated at 5 amps.

**Solution:** We need to use the I/O specifications in Table 5-5 to find the total backplane current required, as follows:

    (a) DCI-24: 200 mA,
    (b) DCO-24: 200 mA,
    (c) AI-4-20MA: 100 mA, and
    (d) AO-4-20MA: 100 mA.

So that, the total backplane current required is

    5 × 200 mA = 1000 mA
    3 × 200 mA =  600 mA
    1 × 500 mA =  500 mA
    1 × 400 mA =  100 ma
                 2200 mA

Therefore, the total backplane current demand is 2200 mA or 2.2 amps. This is less than the rack power supply and backplane current rating of 5 amps, so I/O system current demand is acceptable.

| Model Number | Module Description | Back-Plane Current | I/O Power Rating |
|---|---|---|---|
| AI-4-20MA | Analog input (4-20 mA) | 400 mA | 100 mW |
| AO-4-20MA | Analog output (4-20 mA) | 400 mA | 100 mW |
| ACI-120 | 120 Vac input | 200 mA | 240 W |
| ACI-120 | 220 Vac input | 200 mA | 440 W |
| DCI-12 | 12 volt dc input | 200 mA | 24 W |
| DCI-24 | 24 volt dc input | 200 mA | 48 W |
| DCI-48 | 48 volt dc input | 200 mA | 96 W |
| IACI-120 | Isolated ac input | 200 mA | 240 W |
| ACO-120 | 120 Vac output | 250 mA | 240 W |
| ACO-220 | 220 Vac input | 250 mA | 440 W |
| DCO-12 | 12 volt dc output | 200 mA | 24 W |
| DCO-24 | 24 volt dc output | 200 mA | 48 W |
| DCO-48 | 48 volt dc output | 200 mA | 96 W |
| CC-4NO | 4 NO contact output | 500 mA | 240 W |
| TTLI | TTL input | 150 mA | 100 mW |
| TTLO | TTL output | 150 mA | 100 mW |
| IACO-120 | Isolated ac output | 250 mA | 240 W |

**Table 5-5. Typical Specifications for I/O Modules.**

## Mechanical and Environmental Specifications

The typical mechanical specifications are I/O points per module and wire size. The I/O points per module simply define the number of field points that are controlled or sensed by a module. Typically, modules will have 2, 4, 8, 16, or 32 points per module. The higher density modules require higher operating current, so the backplane current must be carefully checked. The number of wires is also increased and it might be a problem for larger gage wires. The wire size specification defines the number of conductors and the largest wire gage that the I/O terminal points will accept. For example, a typical wire specification for an ac input/output module is 2-14 AWG wires per terminal.

The important environmental specifications are the ambient temperature rating and the humidity rating. The ambient temperature specification defines the maximum temperature of the surrounding air in which the I/O system will operate properly. This rating is based on the heat dissipation characteristics of the circuit components inside the I/O modules, which are much higher than the module ambient temperature rating. A typical value of ambient temperature rating is 0° to 60°C. Exceeding the ambient temperature can be dangerous because the internal circuits of the modules will act erratically, resulting in undesirable outputs to the process being controlled.

The humidity specification is typically 5% to 95% without condensation. The system designer must ensure that the humidity is properly controlled in the plant area where the I/O system is installed.

Adherence to the PLC manufacturer's specifications will ensure proper and safe operation of the control system.

## EXERCISES

5.1 Describe the main functions and advantages of programmable controller I/O equipment racks.

5.2 What are the purposes and functions of the backplane on a typical I/O equipment rack?

5.3 Describe the operation and purposes of the sections of a typical ac input module.

5.4 Draw a wiring diagram for a typical +24 Vdc input module with four temperature high switches (TSH-100, 101, 102, and 103) and three pressure low switches (PSL-210, 211 and 212) connected to the module.

5.5  Draw a wiring diagram for a typical +24 Vdc output module connected to four solenoid valves (TV-100, 101, 102, and 103) and three pump starters (P-100, 200 and 300).

5.6  Describe the operation of the internal circuits in a typical ac output module.

5.7  Draw the wiring diagram for a BCD output module connected to a four-digit LED display. Assume the output of the module is the decimal number 2783 and show the binary bits out to each LED.

5.8  Calculate the backplane current requirements for a 16-slot I/O rack with the following modules installed: (a) four 12 Vdc input modules, (b) six 12 Vdc output modules, (c) four analog input modules, and (d) two analog output modules. Assume the rack power supply and the backplane circuit card are rated at 5 amps.

## BIBLIOGRAPHY

1.  Jones, C. T., and Bryan, L. A., *Programmable Controllers: Concepts and Applications,* International Programmable Controls, Inc., 1983.

2.  Hughes, T. A., *Basics of Measurement and Control,* ISA Publications, Second Edition, 1995.

3.  *Processor Manual PLC-5 Family Programmable Controllers,* Allen-Bradley Co., Inc., 1987.

4.  *Programming and Operations Manual, PLC-2/30 Programmable Controllers,* Allen-Bradley Co., Inc., 1988.

5.  *Allen-Bradley Industrial Computer and Communications Group Product Guide,* Allen-Bradley Co., Inc., 1987.

6.  *Allen-Bradley Automation Systems,* Allen-Bradley Co., Inc., December 1994.

7.  Bryan, L. A., and Bryan, E. A., *Programmable Controllers: Theory and Implementation,* Industrial Text Co., 1988.

# 6

# Memory and Storage Devices

## Introduction

Programmable controller systems store information and programs by either electronic or electromechanical means. Electronic methods, generally called *memory*, are used to store the control program for the programmable controller system, and it is usually located in the same housing as the processor. The information stored in memory determines how the input and output data will be processed by the programmable controller.

However, electronic memory circuits have limited storage capacity and the most common varieties lose data when power is interrupted for even a split second, so there must be electromechanical means to permanently back up control data and instructions. Commonly called storage devices, these electromechanical devices once stored data in the form of holes punched in paper tape or cards; now the holes have been supplanted by infinitesimally small areas of magnetism on floppy or hard disks and magnetic tapes and by pits (indentations) and lands (elevations) on compact disks (CDs).

Storage devices for programmable controllers are electromechanical in the sense that retrieving stored data requires the use of electrical/mechanical machinery such as tape player, CDs, or magnetic disk units in order to locate the data and convert it into electronic pulses acceptable to the programmable controller. Storage devices are slower than electronic memory, but their main advantage is high capacity permanent storage. For example, information or programs recorded on magnetic tape, disks, or CDs can be retained for years

## Memory

Programmable controller memories can be visualized as a two-dimensional array of storage cells, each of which can store a single bit of information in the form of a 1 or a 0. This single "bit" gets its name from BInary digiT. A bit is the smallest structural unit of memory and stores information in the form of 1s and 0s. Ones and zeros are not actually in each cell; each cell has a voltage present at the output of an electronic circuit, indicated by a one, or voltage not present, indicated by a zero.

The bit is set or ON if the stored information is 1, and OFF if the stored information is 0. In most cases, it is necessary for the processor to handle more than a single bit. For example, when transferring data to and from memory, storing numbers, and programming codes, a group of bits called a byte or word is required. A byte is defined as the smallest group of bits that can be handled by the CPU at one time. In programmable controllers, byte size is normally 8 bits and word size is normally 2 bytes or 16 bits but the word size can be smaller or larger depending on the specific microprocessor being used.

Memory capacity is specified in thousands or "K" increments where 1K is 1024 words (i.e., $2^{10} = 1024$) of storage space in most cases. Programmable controller memory capacity may vary from less than one thousand bits to over 64,000 words (64K words), depending on the programmable controller manufacturer. The complexity of the control plan, the number of I/O points, and type of I/O will determine the amount of memory required.

Word length is usually 2 byte (16 bits) or more in length. Typical word lengths in programmable controllers are 8, 16, or 32 bits. A 16-bit word is shown in Figure 6-1.

Many programmable controllers use the octal numbering system to identify each bit in a given word as shown in the figure. The most significant bit (MSB) is bit 17, and the least significant bit (LSB) is bit 00.

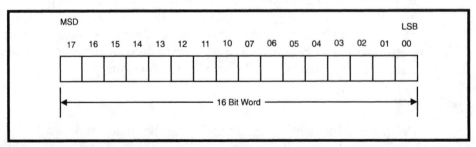

**Figure 6-1. Basic memory word.**

A simple 64-cell memory array is shown in Figure 6-2. This array consists of 8 rows and 8 columns. This 64-bit array requires only 6 bits to address a given cell. A cell is normally an electronic circuit called a flip-flop that can have a value of 5 volts (logic 1) or 0 volts (logic 0). To retrieve data from the memory array, row and column address decoders select the appropriate cell.

These memory arrays are normally provided by integrated circuits (ICs). A typical IC unit contains many thousands of memory cells arranged in various ways. An 8K-bit (8096 cells) memory integrated circuit, for example, could be arranged as 8K memory cells of 1 bit each, or 1K bytes of 8 bits each. The number of groups (bits, bytes, or words) addressed is a function of $2^n$, for example $1K = 2^{10}$, $4K = 2^{12}$, $8K = 2^{13}$, etc. The value n is the number of address bits needed to select each separate group. For 1000 words it is necessary to use 10 bits to address each word of storage, the word size being 8 bits, 16 bits, and 32 bits. For a $1K \times 8$ bit memory the IC would require 10 address bits to select 1K words of memory. A typical $1K \times 8$-bit memory chip is shown in Figure 6-3. The IC has 8 pins for input and output data bits, 10 connection pins for selecting addresses, 2 pins for control signals (chip enable and read/write), and 2 pins for dc power. The two power supply pins are used for $+5$ Vdc and ground, The Read/Write (R/W)

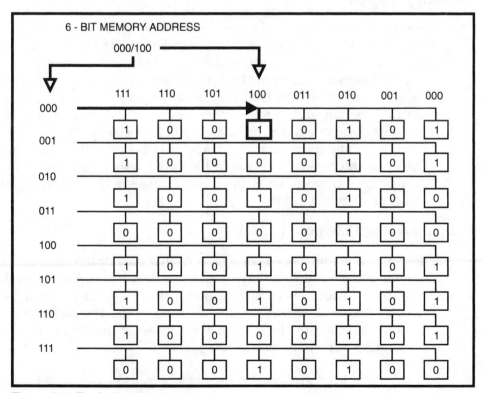

**Figure 6-2. Typical memory array.**

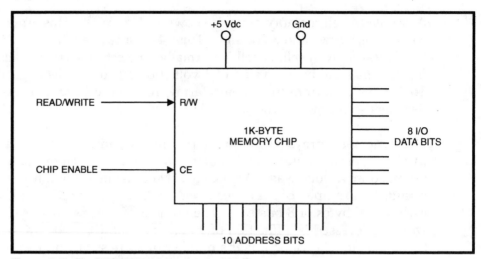

**Figure 6-3. Typical 1K-byte memory chip.**

control signal is used to determine whether the data bits are read into memory (R/W signal low) or data is sent out from memory (R/W is high). The chip enable control signal is used to select operation of each separate chip when a group of integrated circuits is used to provide a larger memory than is provided by only one chip.

# Memory Types

This section will discuss the types of memory generally used in programmable controllers and their applications to the type of data or information stored. In selecting the type of memory to be used, a system designer is concerned with volatility and ease of programming. He is concerned with volatility because memory holds the process control program, and, if this program is lost, production in a plant will be down. Ease in altering the memory is important since the memory is involved in any interaction that takes place between the user and the PLC. This interaction begins with the initial system programming and debugging and continues with on-line changes such as changing timer and counter preset values.

## Random Access Memory (RAM)

Random access memory is designed so that data or information can be written into or read from any unique location. Figure 6-4 shows that data can be placed into RAM memory using the write mode and data can be retrieved from RAM using the read mode. The *address* input to the RAM specifies the location or the address of the data to be read or the location to be written into.

Programmable controllers, for the most part, use RAM with battery backup for application memory. RAM provides an excellent means for

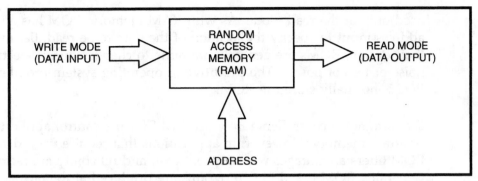

**Figure 6-4. RAM memory block diagram.**

easily creating and altering a control program as well as allowing data entry. In comparison to some other memory types, RAM is relatively fast. The only important disadvantage of battery-supported RAM is that it requires a battery that might fail at a critical time, but most programmable controllers have battery low lights to alert operations personnel to replace the memory power backup battery.

RAM memory is an integrated circuit chip that stores individual bits of data in multiple rows and columns of cells. This row and column arrangement allows each cell to have a unique designation, called an address. This address consists of a row identifier and a column identifier, both expressed as binary numbers as discussed earlier.

## Read-Only Memory (ROM)

Read-only memory is designed to permanently store a fixed program, which normally cannot or will not be changed. It gets its name from the fact that its contents can be read but not written into or altered once the data or program has been stored. Figure 6-5 shows that data can be

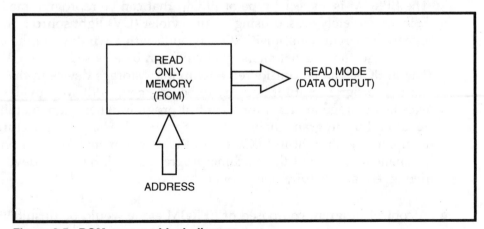

**Figure 6-5. ROM memory block diagram.**

used only in the read mode. As with RAM memory, ROM has an address input to specify the location of the data to be read. Because of their design, ROMs are generally immune to changes due to electrical noise or loss of power. The executive or operating system program of a PLC is normally stored in ROM.

Programmable controllers rarely use ROM for the control applications program memory. However, in applications that require fixed data, ROM offers advantages where speed, cost, and reliability are factors. Generally, ROM-based PLC programs are produced at the factory by the equipment manufacturer. Once the original set of instructions is programmed, it can never be altered by the user. The manufacturer will write and debug the program using a read/write-based controller or computer and then the final program is entered into ROM. ROM application memory is typically found in dedicated programmable controllers.

## Programmable Read-Only Memory (PROM)

The PROM is a special type of ROM that is rarely used in most programmable controller applications. However, when it is used, it will most likely be a permanent storage to some type of RAM. Although PROM is programmable and, like other ROM, has the advantage of nonvolatility, it has the disadvantages of requiring special programming equipment, and, once programmed, it cannot be erased or altered. Any program change would require a new set of PROM chips. PROM memory might be suitable for storing a program that has been thoroughly checked while stored in RAM and will not require further changes or on-line data entry.

## Erasable Programmable Read-Only Memory (EPROM)

The EPROM is a special type of PROM that can be reprogrammed after being completely erased using an ultraviolet (UV) light source. The integrated circuit chip for EPROM is built with a window on the top of the chip so the internal memory circuits can be exposed to the UV light. The EPROM can be considered a temporary storage device in that it stores a program until it is ready to be changed. EPROM provides an excellent storage medium for a control program where nonvolatility is required but program changes are not required. Many manufacturers of equipment with built-in PLCs use EPROM-type memories to provide permanent storage of the machine program after it has been developed, debugged, and is fully operational.

A control program composed of EPROM alone would be unsuitable if online changes and/or data entries are a requirement. However, many

PLCs offer EPROM control program memory as an optional backup to battery supported RAM. EPROM, with its permanent storage capability combined with the easily altered RAM, makes a suitable memory system.

# Storage Units

While integrated circuit memory devices are used for programmable controller internal memory, other devices such as floppy diskettes (disks) or hard disks are used for external programs and data storage. Information is stored on a disk much as music or video is recorded on tape by magnetizing small areas on the disk in a predefined direction. One direction represents a 1 bit, and the opposite direction represents a 0 bit in a microcomputer-based system, such as a programmable controller or a personal computer.

## Storage on Disks

There are three main disk storage systems: the 5.25-inch floppy diskette, the 3.5-inch floppy diskette, and the hard disk drive. Floppy disks and hard drives have similar structures; if we think of a floppy disk as a two-dimensional version of a hard drive, the similarities are very apparent. The floppy disks are small and portable storage mediums but the hard drives have the advantage of very high data storage capacity.

### 5.25-Inch Floppy Disk Drives and Diskettes

A typical 5.25-inch floppy disk consists of two parts: a disk of thin plastic coated with magnetic material and a protective plastic jacket. A typical floppy disk is shown in Figure 6-6. The magnetic coating is

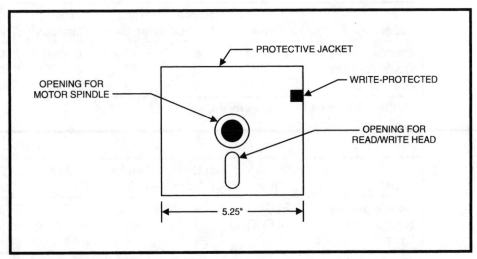

**Figure 6-6.  5.25-inch floppy diskette.**

visible through the opening in the jacket. The hole in the center of the disk goes around the drive motor, which spins the disk so that data can be written or read.

If you cover the write-protect notch with a piece of tape, no data on the disk can be changed. Data is stored on a disk in narrow concentric circles called tracks. These tracks are labeled from 0 to N; N represents the total number of tracks minus 1 and varies depending on the disk read/write format. The outermost track is always numbered 0, the next track is numbered 1, and so on, until the final innermost track is numbered N-1.

Each track is subdivided into a fixed number of sectors and each sector holds the same amount of data. Sectors are numbered from 1 to N; N represents the number of sectors per track. The maximum number of sector in a track depends on the type of floppy disk drive and the diskette's format as listed in Table 6-1 for a 5.25-inch diskette and Table 6-2 for a 3.5-inch diskette.

Each sector contains 512 bytes and is the smallest amount of data that a program can access. The following equation can be used to calculate the capacity of a floppy diskette:

$$Capacity = (sectors) \times (tracks/sector) \times 512 \ bytes/sector$$

This equation is for one side of a diskette. If the floppy disk drive has two read/write heads (like most recent floppy drives), you must multiply the calculated capacity by two.

| Label | Drive | Sectors Per Track | Tracks Per Side | Disk Capacity | Data Transfer |
|---|---|---|---|---|---|
| Double density | PLC/XT | 8 | 40 | 160/320K | 250K/sec |
| Double density | PLC/XT | 9 | 40 | 80/360K | 250K/sec |
| High density | AT | 15 | 80 | 1.2 Meg | 500K/sec |

Table 6-1. 5.25-inch Diskette Formats.

| Label | Drive | Sectors Per Track | Tracks Per Side | Disk Capacity | Data Transfer |
|---|---|---|---|---|---|
| Double density | PLC/XT | 9 | 80 | 720K | 250K/sec |
| High density | PLC/XT | 18 | 80 | 1.44 Meg | 500K/sec |
| Extra high density | AT | 36 | 80 | 2.88 Meg | 1 Mbit/s |

Table 6-2. 3.5-Inch Diskette Formats.

Tracks on a 5.25-inch double density PLC/XT disk are numbered 0 through 39 (i.e., 40 tracks); sectors are numbered 1 through 9, for a total of 360 sectors (9 sectors per track times 40 tracks) on each side. Most personal computers have double-sided disk drives, which use both sides of a disk. A double-sided disk can store 368,640 bytes or 360K bytes.

The high capacity 5.25 disks on the IBM personal computer AT have 80 tracks, each of which has 15 sectors. A sector still holds 512 bytes, so a high capacity disk can store 1,228,800 bytes or 1.2 mega bytes as shown in Table 6-1.

### 3.5-inch floppy disk drives and diskettes

Although 5.25-inch floppy disk drives are still used, they are being largely replaced by the smaller 3.2-inch floppy disk drives. These smaller drives first became popular on the laptop computers. Now, 3.5-inch floppy disk drives are also used on most desktop personal computers. The 3.5-inch disks are preferred because of their convenient size and the sturdiness of their rigid plastic case. The magnetic surface of a 3.5-inch disk is covered by a sliding metal door to protect the data from damage by dust and dirt particles in the air as shown in Figure 6-7.

A 3.5-inch disk has a considerably higher track density than a 5.25-inch disk, especially since the diameter of the magnetic media is smaller. The original 3.5-inch floppy disk drives had a double density format with a capacity of 720K. The newer 3.5-inch disks have a 1.44 Meg capacity. A 1.44 Meg floppy disk is now the standard in personal

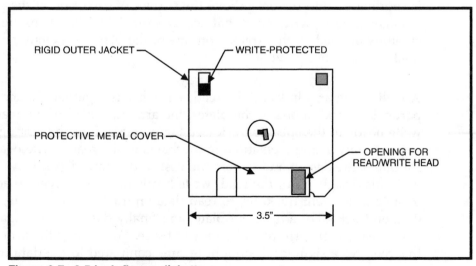

**Figure 6-7. 3.5-inch floppy diskette.**

computer applications. It also has 80 tracks per side, but has 18 sectors per track. It also has double the data transfer rate, which makes it impossible to use these disks in the double density 3.5-inch floppy drives of earlier personal computers.

Recently, a new 3.5-inch format was introduced and disks using this format have a capacity of 2.88 Meg. This format is called extra high density (ED). ED disks have double the number of sectors (36) and can only be read by the new ED floppy disk drives. Like high density drives, ED disk drives are downwardly compatible. This means they are able to read and write to both double density and high density disk formats.

## Disk Drives and Controllers

A 3.5-inch floppy disk drive consists of a motor that rotates the disk at 300 revolutions per minute and a mechanism for moving the read/ write head. The drive also has an electronic circuit or component, called a data separator. This data separator converts a voltage signal into a binary data stream as the read/write head passes over the surface of the disk. The floppy drive is controlled by a disk controller, which is either part of the personal computer's motherboard or on an I/O card in one of the computer's expansion slots.

## Hard Drive Storage

A hard drive is simply a group of magnetic storage plates stacked on top of each other as shown in Figure 6-8. Each magnetic plate is similar to a floppy disk; it has two sides, is divided into tracks, and each track is subdivided into sectors. Above the surface of each side of the plate is a magnetic read/write head that accesses the data. The plates are carefully aligned so that track 0 on one of the plates is exactly above track 0 of another plate.

A read/write arm links all the read/write heads together. To access a particular track on one of the plates, the arm moves all of the read/ write heads to the specific track (see Figure 6-8). Since this arrangement requires only a single positioning mechanism (the read/write head), it simplifies the design and lowers the cost of the hard drive. However, with this design, all of the read/write heads must be moved to access data on a different track. So, to read data on track 1 of one plate, then data on track 40 of a different plate, and finally data on track again, the entire read/write arm must be moved twice. Positioning the arm like this requires a significant amount of time compared to the data transfer time.

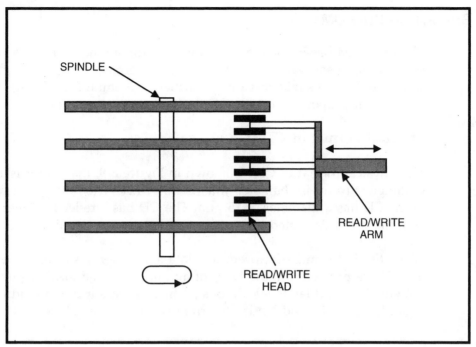

**Figure 6-8. Hard drive structure.**

A hard drive rotates ten times faster than a floppy disk. so its data transfer rate is at least ten times faster than a floppy disk. The data transfer rate increases by a second power of ten, because 3.5-inch hard drives can store almost 100 sectors per track.

## Storage on Tape

Information is stored magnetically on tape in straight line tracks. On a tape, related data bytes are grouped into what is termed a record and stored together in one or more blocks along the tape. Because the read/write head cannot jump from place to place on the tape, locating a needed piece of information may require scanning hundreds of inches of tape. In some cases, a read operation can take over 30 seconds to complete. Since this time period is normally unacceptable in process or machine control operations, tape units are normally only used to back up programmable controller programs or to archive process data.

The most common tape format is nine track. In this format, the eight bits of each byte of data are recorded simultaneously by eight heads across separate tracks in a line perpendicular to the edge of the tape. The remaining track is used for the parity bit, which is recorded by a separate head.

## Storage on CD-ROM

The CD-ROM has become a very popular storage medium for both PLC programming software and for Graphical Interface Unit (GUI) software. These applications take advantage of the large amount of storage space and special properties the CD-ROM offers over conventional drives.

### Physical Format of CDs

The physical format of CDs is shown in Figure 6-9, the platter is 120 millimeters (mm) in diameter and 1.2 mm thick. The hole in the center of the CD has a diameter of 15 mm. The CD has a reflective layer of aluminum that is coated by a protective layer of clear paint.

When the CD is manufactured, the data and programs to be stored on the CD are pressed into the layer of aluminum in the form of pits (indentation) and lands (elevations), which represent the individual data bits. The pits and lands are arranged along a single spiral that

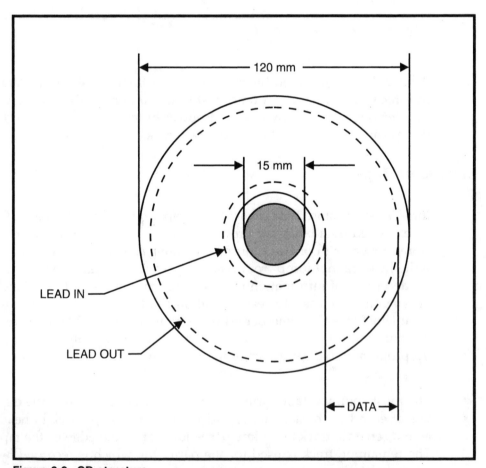

**Figure 6-9. CD structure.**

covers the entire CD, winding from the inside of the disk to the outside. Because the pits are only 0.6 micron (0.6 millionth of a meter) wide, the path this spiral takes is only separated by the microscopically small distance of 1.6 microns. The track density is almost 16,000 tracks per inch (TPI) and this compares to 135 TPI on a 3.5-inch high density floppy diskette. If the data spiral on a CD were stretched out in a straight line it would be approximately 3.75 miles (6 kilometers) long. It also includes no fewer than 2 billion data pits. Naturally, the laser beam used to read the lands and pits of data must be correspondingly small. The scanning laser beam is approximately one micron in diameter which makes it only a little larger than the wavelength of the light that forms its beam.

The active surface of a CD is divided into three sections: the lead-in, the data area, and the lead-out as shown in Figure 6-9. The lead-in area occupies the first four millimeters of the CD's inner edge and contains the table of contents for the stored data and programs. The lead-in is followed by the data area, which can occupy up to 33 millimeters in diameter, depending on the amount of data and programs stored on the CD. Finally, the lead-out section marks the end of the stored data. It follows immediately after the data section and is approximately 1 mm wide.

## Methods of Storing Disk Data

There are two methods used to store data on rotating mass storage systems: constant angular velocity (CAV) and constant linear velocity (CLV). The names of both these methods refer to the rotational speed of the storage medium. Floppy disks and hard drives that are divided into individual tracks and sectors use the CAV method. Regardless of where the disk drive read/write head is located above the disk, the disk always rotates below the read/write head at a constant speed. If the read/write head is above a track at the inner edge of the medium, it travels a much shorter path than it would over an outer track. Hard drives take advantage of this factor by packing more sectors into the larger areas of the outer tracks.

The most important difference between the CAV and the CLV methods is the rotational speed. The rotational speed of the medium does not change with the CAV method, and this is true regardless of the location of the read/write head. The opposite is true with the CLV method used by Cds. The read/write head on a CD always travels a constant distance in a specific unit of time regardless of whether the head is at the inner or outer edge of the medium. However, the rotational speed must be changed based on the position of the head. Therefore, the rotational speed of the drive increases as the head moves from the inner edge of the CD medium to its outer edge. This is one reason why a CD-ROM unit has much slower access times than a hard drive. A CD drive must constantly change its rotational speed. The time to speed up and to slow down becomes significant. Another reason for the slow access time on CDs is that it is much more difficult to find data along a 3.75 mile long spiral than it is to find the same data on a disk that is neatly organized into tracks and sectors on a hard drive.

## Storage of Bits and Bytes on CD

Since all data on a CD is contained on a single continuous spiral, it consists of a single data channel. So bits on a CD are referred to as "channel bits." The transition from a land to a pit or the transition from a pit to a land represents a binary one. Lands and pits are used to represent a binary zero. The length of a land or the length of a pit determines how many binary zeros are represented. This is basically the same procedure used to record data on magnetic storage devices like hard disks and floppy disks. The only difference is that magnetic flux changes replace the pits and lands.

Due to technical manufacturing limitations, the minimum length of either a pit or a land is 3 bits and the maximum length is 11 bits. There is a problem trying to represent two consecutive "1" bits when the technical limitations require no less than two and no more than ten binary ones between transitions. Using this scheme, it is not possible to represent all possible combinations of ones and zeros. Instead, a method called eight-to-fourteen modulation (EFM) is used. Using EFM, an 8-bit binary byte that is to be stored on a CD is converted into a 14-bit binary byte. A partial listing of the EFM binary code is shown in Table 6-3.

| Decimal | Binary | EFM Code |
|---------|--------|----------|
| 0 | 00000000 | 01001000100000 |
| 1 | 00000001 | 10000100000000 |
| 2 | 00000010 | 10010001000000 |
| 3 | 00000011 | 10001000100000 |
| 4 | 00000100 | 01000100000000 |
| 5 | 00000101 | 00000100010000 |
| 6 | 00000110 | 00010000100000 |
| 7 | 00000000 | 00100100000000 |
| 8 | 00000000 | 01001001000000 |
| 9 | 00000000 | 10000001000000 |
| 10 | 00000000 | 10010001000000 |

**Table 6-3. EFM Binary Code (Partial Listing).**

The complete EFM Code conversion table is used by the CD-ROM drive control unit to convert the stored EFM data into standard binary 8 bit binary code for use by a personal computer. The codes in the EFM table are selected to avoid two adjacent 1 bits and to stay within the maximum length of eleven 0 bits. However, even conversion by the EFM table does not consider the separation of individual bytes. When the first byte ends in a one bit and therefore changes from a land to a pit or vice versa, the next byte cannot again start with the same change because there is no room between the two bytes. To correct the problem, each byte with its 14 bits is extended by three additional bits called *merge bits*. Merge bits separate the bytes from each other and increase the bits to 17 per byte on the CD.

# Memory Organization

There are two main parts of the programmable controller memory: the system memory and the application memory. The *system memory* is a permanently stored collection of programs that is the operating system for the PLC. This operating system directs activities such as execution of the control program, communication with the peripheral devices, and other system housekeeping functions. The *application memory* consists of the I/O image tables, data registers, and the control program.

The storage and retrieval requirements are not the same for the system operational memory and the application memory. The system memory contains the instructions to operate the CPU. The application memory contains the I/O image table, the control program, and the data table. They do not use the same types of memory. The system memory requires a memory that permanently stores its contents and cannot be deliberately or accidentally altered by loss of electrical power or by the user; therefore, some type of ROM would be used. On the other hand, the user would need to alter the control program and/or the input/output data for any given application, so some type of RAM unit would be used in the I/O image table, the control program, and the data table areas.

## Memory Size

The size of memory is an important factor in designing programmable controller-based control systems. Specifying the correct memory size can save hardware costs and avoid lost time later. Proper calculation of memory size means avoiding the possibility of purchasing a PLC that does not have adequate capacity or that is not expandable.

Memory size is normally expandable to some maximum point in most controllers, but it is not expandable in some of the smaller PLCs.

(Smaller PLCs are defined in this context as units that control 10 to 64 input/output devices.) Programmable controllers that handle 64 or more I/O devices are usually expandable in increments of 1K, 2K, 4K, etc. (Recall that each K represents 1024 or $2^{10}$ word locations in memory.) The memory in the larger controllers is generally 64K or higher.

The stated memory size of a PLC is only a rough indication of the memory space available to the user, since some of the memory is used by the controllers for internal functions. Another problem is that there are generally two different word sizes—8-bit and 16-bit—used by PLC manufacturers.

The main problem with determining memory size for an application is that the complexity of the control program is not determined until after the equipment is purchased. However, we generally know the number of I/O points in the system before the hardware is procured. Memory size can be estimated from this fact; a good rule is to multiply the number of I/O points by 10 words of memory. For example, if the system has 100 I/O points, the program will generally be equal to or less than 1000 words. We should keep in mind that program size is affected by the sophistication of the control program. If the application requires data handling or complex control algorithms, such as PID control, then additional memory will be required.

After the system designer determines the minimum memory required for an application, he will normally add an additional 25% to 50% more for program changes, modification, or future expansion.

---

**EXAMPLE 6-1**

**Problem:** Determine memory size needed for a programmable controller system with 173 input points and 125 outputs (assume 25% spare memory capacity).

**Solution:** Total I/O points $= 173 + 125$
$= 298$ points memory required
$= 298$ points $\times$ 10 words/point $+ 25\%$
$= (2980 + 2980 \times 0.25)$ words
$\cong 4023$ words $= 4$ K

To accurately size memory, we need to understand the overall organization of PLC memory.

---

## Input Image Table

The input image table is an array of bits that store the status of discrete inputs from the process, which are connected to input modules. The number of bits in the table is equal to the maximum number of inputs. A controller with a maximum of 64 inputs would require an input table of 64 bits. Each connected input has a bit in the input table that corresponds exactly to the terminal to which the input is connected. If the input is ON, its corresponding bit in the table is ON (1). If the input is OFF, the corresponding bit is cleared or turned OFF (0).The input table is continuously being changed to reflect the current status of the connected input devices. This status information is also being used by the control program.

The input table shown in Figure 6-10 is the Allen-Bradley PLC5 system of addressing I/O that stores field input data starting at word I:000 and ending at word I:177. In this system, the leftmost letter "I" indicates an input to the PLC, the second digit gives the I/O rack number (racks 1 through 7), and the final digit gives the module group number in the I/O rack.

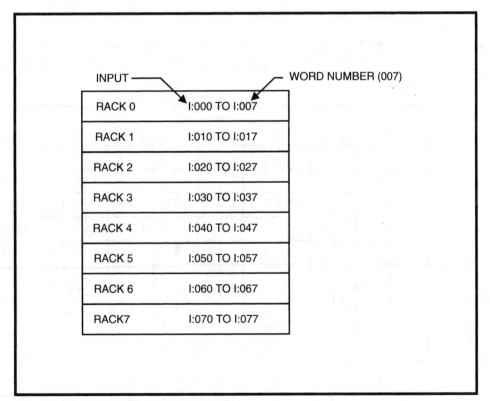

**Figure 6-10. A-B PLC5 input image table.**

Figure 6-11 shows a typical example of a single input bit in an input image table. In this example, input point I:007/12 is identified on the memory map.

## Output Image Table

The output table is an array of bits that controls the status of discrete output devices, which are connected to output interface circuits. The number of bits in the output table is equal to the maximum number of outputs. For example, a programmable controller with a maximum of 512 outputs would require an output image table of 512 bits.

The output image table shown in Figure 6-12 for Allen-Bradley PLC5s uses the same numbering system as the input image table, except the leftmost letter is "O" to signify an output, the first two digits indicates the I/O rack number (racks 00 through 07), and the final digit gives the module group number in the I/O rack (0 through 7).

Each connected output has a bit in the output table that corresponds exactly to the terminal to which the output is connected. Note that the bits in the output table are controlled by the CPU as it interprets the control program and are updated accordingly during the I/O scan. If a bit is turned ON (1), then the connected output is switched ON. If a bit is cleared or turned OFF (0), the output is switched OFF.

Figure 6-13 shows a typical example of a single output bit in an output image table. In this example, output O:017/16 is shown on the memory map.

| | | | | | | | BIT NUMBER | | | | | | | | | WORD |
|---|---|---|---|---|---|---|---|---|---|---|---|---|---|---|---|---|
| 17 | 16 | 15 | 14 | 13 | 12 | 11 | 10 | 07 | 06 | 05 | 04 | 03 | 02 | 01 | 00 | NUMBER |
| 0 | 0 | 0 | 0 | 0 | 0 | 0 | 0 | 0 | 0 | 0 | 0 | 0 | 0 | 0 | 0 | I:000 |
| 0 | 0 | 0 | 0 | 0 | 0 | 0 | 0 | 0 | 0 | 0 | 0 | 0 | 0 | 0 | 0 | I:001 |
| 0 | 0 | 0 | 0 | 0 | 0 | 0 | 0 | 0 | 0 | 0 | 0 | 0 | 0 | 0 | 0 | I:002 |
| 0 | 0 | 0 | 0 | 0 | 0 | 0 | 0 | 0 | 0 | 0 | 0 | 0 | 0 | 0 | 0 | I:003 |
| 0 | 0 | 0 | 0 | 0 | 0 | 0 | 0 | 0 | 0 | 0 | 0 | 0 | 0 | 0 | 0 | I:004 |
| 0 | 0 | 0 | 0 | 0 | 0 | 0 | 0 | 0 | 0 | 0 | 0 | 0 | 0 | 0 | 0 | I:005 |
| 0 | 0 | 0 | 0 | 0 | 0 | 0 | 0 | 0 | 0 | 0 | 0 | 0 | 0 | 0 | 0 | I:006 |
| 0 | 0 | 0 | 0 | 0 | 1 | 0 | 0 | 0 | 0 | 0 | 0 | 0 | 0 | 0 | 0 | I:007 |

INPUT I:007/12

**Figure 6-11. A-B PLC5 input bit example.**

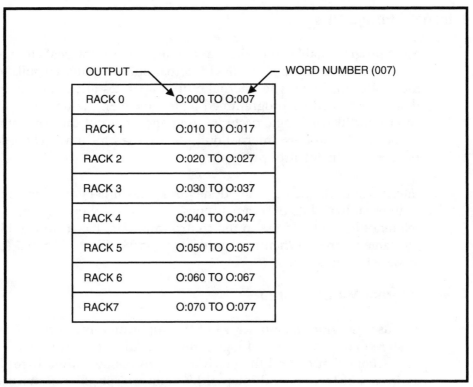

**Figure 6-12. A-B PLC5 output image table.**

| | BIT NUMBER | | | | | | | | | | | | | | | WORD |
|---|---|---|---|---|---|---|---|---|---|---|---|---|---|---|---|---|
| 17 | 16 | 15 | 14 | 13 | 12 | 11 | 10 | 07 | 06 | 05 | 04 | 03 | 02 | 01 | 00 | NUMBER |
| 0 | 1 | 1 | 0 | 0 | 0 | 1 | 0 | 0 | 1 | 0 | 0 | 0 | 0 | 1 | 0 | O:000 |
| 0 | 1 | 0 | 0 | 0 | 1 | 1 | 0 | 0 | 1 | 1 | 0 | 0 | 0 | 1 | 0 | O:001 |
| 0 | 1 | 1 | 0 | 0 | 0 | 1 | 0 | 0 | 0 | 1 | 0 | 0 | 0 | 0 | 0 | O:002 |
| 0 | 0 | 1 | 0 | 0 | 0 | 1 | 1 | 0 | 0 | 1 | 0 | 1 | 0 | 1 | 0 | O:003 |
| 0 | 1 | 1 | 0 | 1 | 0 | 1 | 0 | 0 | 1 | 1 | 0 | 0 | 0 | 1 | 1 | O:004 |
| 0 | 1 | 0 | 0 | 0 | 1 | 1 | 0 | 0 | 0 | 1 | 0 | 0 | 0 | 0 | 0 | O:005 |
| 0 | 1 | 0 | 0 | 0 | 1 | 0 | 1 | 0 | 1 | 1 | 0 | 0 | 0 | 1 | 0 | O:006 |
| 0 | 1 | 1 | 0 | 0 | 0 | 1 | 0 | 0 | 0 | 1 | 0 | 0 | 1 | 1 | 0 | O:007 |

OUTPUT O:007/02

**Figure 6-13. A-B PLC5 output bit example.**

## Internal Storage Bits

Most programmable controllers assign an area for internal storage bits. These storage bits are also called internal outputs, internal coils, or internal control bits. The internal output operates just as any output that is controlled by programmed logic; however, the output is used strictly for internal logic programming and does not directly control an output to the process. Internal outputs are used for interlocking logic purposes in the control program.

Internal outputs include the "done" bits on counters and timers as well as internal logic bits of various types. Each internal output bit, referenced by an address in the control program, has a storage bit of the same address. When the control logic is TRUE, the internal (output) storage bit turns ON.

## User Program Memory Area

The user program memory area of the application memory is used to store the process control logic program. All of the controller instructions that control the machine or process are stored here. The addresses of the real and internal I/O bits are specified in this section of memory. When the PLC is in the RUN mode and the control program is executed, the CPU interprets these memory locations and controls the bits in the data table, which corresponds to a real or internal I/O bit. The interpretation of the control program is accomplished by the processor's execution of the control program.

The maximum amount of user program memory available is normally a function of the controller memory size. In medium and large programmable controllers, the user program size is normally flexible, through altering the data table size so that it meets the minimum data storage requirements. In small PLCs, however, the user program size is normally fixed.

# Hardware-to-Software Interface

Probably the most important thing to understand about programmable controllers is how process data, sensed by the input modules, is used by the processor to activate output devices to control the process. This hardware-to-software interface occurs in the input/ output image tables and was briefly discussed earlier. The instruction address is what connects the software to the hardware. We will present a typical addressing method used in programmable controller systems. The addressing scheme to be presented is for the Allen-Bradley (A-B) PLC5 family of controllers.

The A-B PLC5 I/O addressing method uses 6 position codes (a:bbc/dd) to reference both an I/O image table address and a hardware location. In this system, the leftmost position (a) is a I for an input and a O for an output. The next two bits (bb) are the rack number. The next bit (c) is the module group number (0-7). The remaining two digits represent the bit address in the I/O image table word and the terminal number in the I/O module (see Table 6-4).

For example, the address I:001/07 indicates an input device connected to rack 00 and module group 1 at terminal 07 and the address O:074/10 indicates an output device connected to rack 07 at terminal 10 in module group 4.

A typical example will be used to illustrate this addressing method.

---

**EXAMPLE 6-2**

**Problem:** List the memory bit address for an input connected to an A-B PLC-5/15 rack 3, module group 4, terminal 10.

**Solution:** I:034/10

---

A hardware-to-software interface for an A-B PLC5/15 application is illustrated in Figure 6-14 and shows the operational relationship between the field devices, the input/output image table, and the user ladder logic program.

In the example shown in Figure 6-14, if the high level switch connected to an input module in rack 0, module group 0, and terminal 7 is closed, the internal software bit I:000/07 will be set to 1. The dotted line from terminal 7 to memory bit location I:000/07 indicates a connection within the PLC system. If, at the same time, the valve open position switch that is connected to terminal 13 of the same input module is

| a | I/O address identifier:   I = input device<br>O = output device | |
|---|---|---|
| bb | I/O Rack number:<br>PLC-5/10, -5/12, -5/15, -5/20, 20E<br>PLC-5/25, -5/30<br>PLC-5/40, -5/40E, -5/40L | 00-03 (octal)<br>00-07 (octal)<br>00-17 (octal) |
| c | I/O Group number 0-7 (octal) | |
| dd | Terminal (bit) number 00-17 (octal) | |

**Table 6-4. A-B PLC5 I/O Addressing Scheme (a:bbc/dd).**

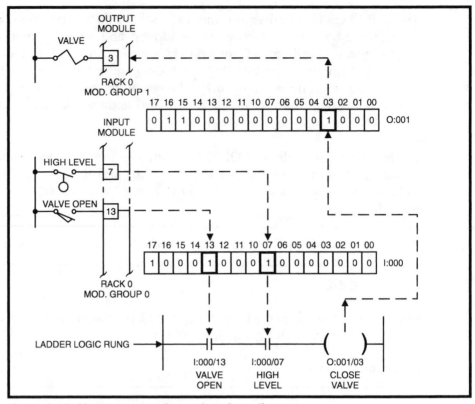

**Figure 6-14. Hardware-to-software interface diagram.**

closed, then input bit I:000/13 will be set to 1 and the ladder logic rung will have logic continuity and output bit O:001/03 will be set to 1. The PLC will engerize the output circuit connected to terminal 3 of output module in Rack 0 and module group 1. This in turn will energize the solenoid valve connected to terminal 3.

The ladder logic rung in Figure 6-14 shows an example of two external inputs being used to set an external output bit. We can also have ladder logic rungs that use internal bits to set both external and internal bits. In some cases, the logic in the rung will use a combination of both internal and external bit to control an output bit.

## EXERCISES

6.1 Describe the function and purpose of the two main parts of programmable controller memory.

6.2 List the advantages and disadvantages of the following memory types: RAM, ROM, PROM, and EPROM.

6.3 Calculate the storage capacity of a double-sided floppy disk with 36 sectors per track and 80 tracks per side.

6.4    Discuss the method used to store data on magnetic disks.

6.5    Explain the basic operation of disk drives and controllers.

6.6    Discuss the basic operation of hard disk storage devices.

6.7    Explain the two methods used to store data on rotating mass storage systems.

6.8    Determine memory size needed for a programmable controller system with 235 input points and 195 outputs (assume 20% spare memory capacity).

6.9    List the memory bit addresses for the following I/O points in an Allen-Bradley PLC5 system:
(a) an input to rack 1, module group 3, terminal 1.
(b) an output from rack 2, module group 1, terminal 7.
(c) an input to rack 2, module group 2, terminal 13.

6.10   Describe the functions of the system memory and the application memory in a PLC system.

## BIBLIOGRAPHY

1.    Gilbert, R. A., and Llewellyn, J. A., Programmable *Controllers: Practices and* Concepts, Industrial Training Corporation (ITC), 1985.

2.    Jones, C. T., and Bryan, L. A., Programmable *Controllers: Concepts and Applications*, International Programmable Controls, Inc., 1983.

3.    Hughes, T. A., *Basics of Measurement and Control*, Instrument Society of America, Second Edition, 1995.

4.    *Processor Manual PLC-5 Family Programmable Controllers*, Allen-Bradley Co., Inc., 1987.

5.    *Programming and Operations Manual, PLC-2/30 Programmable Controllers*, Allen-Bradley Co., Inc., 1988.

6.    *Understanding Computers: Memory and Storage*, Time-Life Books, Inc., 1987.

7.    Boylestad, R. L., and Nashelsky, L., *Electronic Devices and Circuit Theory*, Third Edition, Prentice-Hall, Inc., 1982.

# 7

# Basic PLC Programming

## Introduction

Programming language allows the user to communicate with the programmable logic controller (PLC) via a programming device. PLC manufacturers use several different programming languages, but they all convey to the system, by means of instructions, a basic control plan.

The five most common types of languages encountered in programmable controller system design are structured text (ST), instruction list (LT), ladder diagram (LD), function block diagrams (FBDs), and sequential function chart (SFC). These languages can be grouped into two major categories: textual languages and graphical languages. The first two, ST and IL, are textual languages, whereas the other three, LD, FBDs, and SFC, are graphical languages.

The basic programmable controller languages consist of a set of instructions that will perform the most basic type of control functions: relay replacement, timing, counting, sequencing, and logic. However, depending on the controller model, the instruction set may be extended or enhanced to perform other operations. The need for these additional functions have been brought about by a need to execute more powerful instructions that go beyond the simple timing, counting, and ON/OFF control. These additional functions are used for analog control, data manipulation, reporting, complex control logic, and other functions that are not possible with the basic instruction sets of LD and LT.

The language used in a PLC actually dictates the range of applications in which the controller can be applied. Depending on the size and capabilities of the controller, one or more languages may be used in combination. Typical combinations of languages are:

1. Ladder diagram (LD);
2. Instruction list (IL);
3. Sequential function chart with LD, IL, ST, and/or FBD;
4. Function block diagrams; and
5. Structured text.

The most commonly used PLC language is ladder diagram and it will be discussed in this chapter. We will also discuss the documentation of PLC programs. But first, we will cover PLC programming standards.

# PLC Programming Standards

There have been numerous PLC programming standards proposed by different national and international committees to develop a common interface for programmable controllers.

## Brief History of PLC Standards

In 1979, a working group of international PLC experts was appointed by various national committees to write a first draft of a comprehensive PLC standard. The first committee draft was issued in 1982.

After an initial review of the document by the national committees, it was decided that the standard was too complex to handle as a single document. As a result, the working group was split into five task forces, one for each part of the standard. The subject of each part is as follows: Part 1, General Information; Part 2, Equipment and Testing Requirements; Part 3, Programming Languages; Part 4, User Guidelines; and Part 5, Communications. Each task group consists of several international experts, each backed by a national advisory group.

IEC 1131-3, the standard for PLC programming languages, was issued by the IEC in March of 1993.

## IEC 1131-3 Standard Languages

The IEC 1131-3 standard has three graphical languages—ladder diagram (LD), function block diagram (FBD), and sequential function chart (SFC)—and two text-based languages—instruction list (IL) and structured text (T). The standard allows different parts of an application to be programmed in different languages that can be combined into a single executable program.

LD uses a standardized set of ladder logic symbols that basically are representations of relay ladder diagrams; it is the most widely used.

FBD has been widely used in Europe. Its programming symbols appear as blocks connected together like a circuit diagram; this makes it well suited for many applications involving the flow of information or data between control components.

SFC is a graphical language used to describe sequential operations. The control process is represented as a set of well-defined steps, linked by transitions. A Boolean logic condition is attached to each transition. Actions within the steps are detailed by using the other languages (LD, FDB, ST, and IL).

IL is a low level language. It is very effective for small simple applications or for optimizing parts of an application. Instructions always relate to the current result and the operator indicates the operation that must be made between the current value and the operand. The result of the operation is stored again in the current result.

ST is a high level structured language designed for automation processes. This language is used mainly to implement complex procedures that cannot be easily expressed with graphical languages. ST is the default language for the description of the actions within steps and conditions attached to the transitions of the SFC language.

## Ladder Diagram Language

LD language is a symbolic instruction set that is used to create a programmable controller program. It is composed of six categories of instructions that include relay type, timer/counter, data manipulation, arithmetic, data transfer, and program control. The ladder instruction symbols can be formatted to obtain the desired control logic that is to be entered into memory.

The main function of the LD program is to control outputs based on input conditions. This control is accomplished through the use of what is referred to as a ladder rung. Figure 7-1 shows the basic structure of a ladder logic program. In general, a logic rung consists of a set of input conditions represented by relay contact-type instructions and an output instruction at the end of the rung represented by the coil symbol. Throughout this section, the contact instructions for a rung may be referred to as input conditions, rung conditions, or control logic.

Coils and contacts are the basic symbols of the ladder diagram instruction set. The contact symbols programmed in a given rung

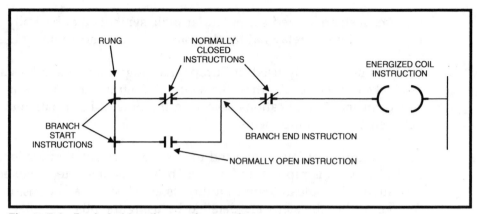

**Figure 7-1. Basic ladder diagram instructions.**

represent conditions to be evaluated in order to determine the control of the output; all outputs are represented by coil symbols.

When programmed, each contact and coil is referenced with an address number that identifies what is being evaluated and what is being controlled. Recall that these address numbers reference the data table location of either an internal bit or a connected input or output. A contact, regardless of whether it represents an input/output connection or an internal bit, can be used throughout the program whenever that condition needs to be evaluated.

The format of the rung contacts is dependent on the desired control logic. Contacts may be placed in configurations such as series, parallel, or series and parallel that are required to control a given output. For an output to be activated or energized, at least one left-to-right path of contacts must be closed. A complete closed path is referred to as having logic continuity. When logic continuity exists in at least one path, it is said that the rung condition is TRUE. The rung condition is FALSE if no path has continuity.

In the early years, the standard ladder instruction set was limited to performing only relay equivalent functions using the basic relay-type contact and coil symbols similar to those illustrated in Figure 7-1. A need for greater flexibility, coupled with developments in technology, led to extended ladder diagram instructions that perform data manipulation, arithmetic, and program flow control. We will discuss relay-type instructions first and the more advanced instructions later.

## Relay-Type Instructions

The relay-type instructions are the most basic of programmable controller instructions. They provide the same capabilities as hardwired relay logic discussed in Chapter 4, but with greater flexibility. These instructions primarily provide the ability to examine the ON/OFF status of a specific bit addressed in memory and to control the state of an internal or external output bit. The following is a description of relay-type instructions that are most commonly available in any controller that has a ladder diagram instruction set.

## Normally Open Instruction

The normally open instruction is programmed when the presence of the input signal is needed to turn an output ON. When evaluated, the referenced address is examined for an ON (1) condition. Allen-Bradley (A-B), calls this type of instruction "Examine On" and uses the mnemonic XIC in their programming software. The referenced address may represent the status of an external input, external output, or internal output. If, when examined, the referenced address is ON, then the normally open instruction will close and allow logic continuity (logic flow). If it is OFF (0), then the normally open instruction will assume its normal programmed state (open), thus breaking logic continuity.

## Normally Closed Instruction

The normally closed instruction is programmed when the absence of the referenced signal is needed to turn an output ON (1). When evaluated, the referenced address is examined for an OFF (0) condition. A-B calls this type of instruction "Examine Off" and uses the mnemonic XIO to represent it in developing programs. The referenced address may represent the status of an external input, external output, or an internal output. If, when examined, the referenced address is OFF, then the normally closed instruction will remain closed, allowing logic continuity. If the referenced address is ON, then the normally closed contact will open and break logic continuity.

## Branch Start Instruction

The branch start instruction begins each parallel logic branch of a rung. It is the first instruction programmed if a parallel branch or logical OR function is needed in a logic rung. For example, if we need to perform the logical OR of input A (I:001/00) and input B (I:001/07) we would begin with a branch start as shown in Figure 7-2.

## Branch End Instruction

The branch end instruction finishes a set of parallel branches. This instruction is used after the last instruction of the last branch to

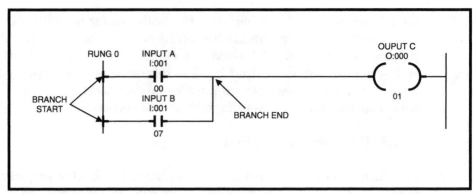

**Figure 7-2. Branch start and end instructions.**

complete a set of parallel branches. To complete the continuity path for the OR circuit of Figure 7-2, we need to add a branch end instruction as shown.

### Energize Coil Instruction

The energize coil instruction is programmed to control either an output connected to the controller or an internal output bit. If any rung path has logic continuity, the referenced output is energized or ON. Many PLC manufacturers call this type of instruction "output energized" and use the mnemonic OTE. The output bit is turned OFF if logic continuity is lost. When the output is ON, a normally opened instruction of the same address will close, and a normally closed instruction will open. In Figure 7-2, the output instruction O:000/01 is energized or true if either input A or input B is true or if both inputs are true.

Several example applications will help to illustrate ladder diagram programming.

---

**EXAMPLE 7-1**

**Problem:** Write a ladder diagram program to start and stop a pump. In this application, the normally open (NO) contacts of a local panel-mounted start pushbutton are wired to input I:001/01, and the normally closed contacts of a stop pushbutton are connected to input I:001/00. The pump starter relay is connected to PLC output O:003/01 and the auxiliary (aux.) motor start contacts (NO) are connected to PLC input I:001/02.

**Solution:** The application can be performed using the LD program of Figure 7-3.

When the NO start pushbutton is depressed, input I:001/01 is true and since the NC stop pushbutton is not depressed, input I:001/00 is also true, so that there is logic continuity in Rung 0, and the output bit O:003/01 is energized. This output energizes the pump start relay and the pump auxiliary switch is closed. This, in turn, sets bit I:001/02 ON to seal-in the start pushbutton (PB) and hold the pump ON until the stop PB is depressed. After the stop PB is depressed, the pump will be turned off and the auxiliary contacts will open and set input bit I:001/02 to false.

It is important to note that Stop pushbuttons for moving equipment always use normally closed (NC) contacts for safe operation. The NC contacts are used so that if a wire from the PB to the PLC is cut or removed the moving equipment cannot be started. If the NC contacts on the Stop PB are used and the control wire from the PB to the PLC is cut or removed when the equipment is energized the moving equipment, such as a pump or motor, will be turned off.

---

**EXAMPLE 7-2**

**Problem:** There is only a single field-mounted pushbutton available to turn an electric warning beacon on and off. Write a ladder program to control the beacon. Assume the normally open contacts of the push button are wired to input point I:001/01, and the beacon is connected to output O:003/00.

**Solution:** The warning beacon can be controlled using the LD program of Figure 7-4.

---

In this example program, when the beacon control pushbutton (PB) is depressed for the first time, output bit (O:003/00) is energized and it turns ON the warning beacon. This output control bit (O:003/00) also seals itself in. If the pushbutton is depressed again, the beacon is turned off. The second rung (rung number 1) of the program detects the first time the pushbutton is depressed, while the first rung (rung number 0) senses the second time the pushbutton is depressed. The last rung

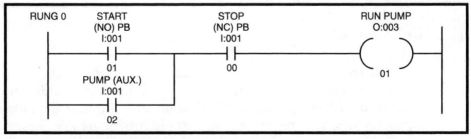

**Figure 7-3. Pump start/stop LD program.**

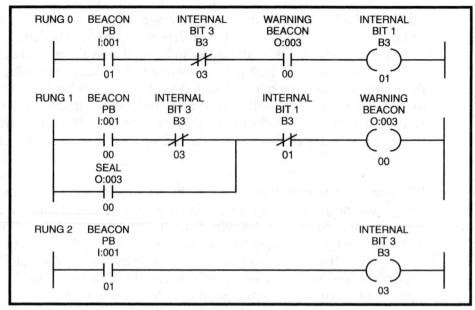

**Figure 7-4. Alarm beacon application program.**

(rung 2) is used to control internal bit 3 (B3/03). The normally closed contacts of internal bit 3 are used in rungs 0 and 1 to help perform the "push to start and push to stop" function in the ladder diagram program.

### Latch Coil

The latch coil instruction is programmed, if it is necessary, for an output to remain energized even though the status of the input bits that caused the output to energize may change. If any rung path has logic continuity, the output is turned ON and retained ON even if logic continuity or system power is lost. The latched output will remain latched ON until it is unlatched by an output instruction of the same reference address. The unlatch instruction is the only automatic (programmed) means of resetting the latched output. Although most controllers allow latching of internal or external outputs, some are restricted to latching internal outputs only.

### Unlatch Coil

The unlatch coil instruction is programmed to reset a latched output of the same reference address. If any rung path has logic continuity, the referenced address is turned OFF. The unlatch output is the only automatic means of resetting a latched output. Figure 7-5 illustrates the use of the latch and unlatch coils to start or stop a pump.

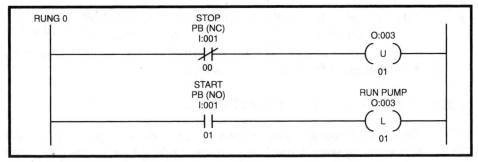

**Figure 7-5. Latch and unlatch pump start/stop program.**

In this example, the NC contacts of the start PB and the NC contacts of the Stop PB are connected to discrete PLC inputs. The NC contacts of the stop PB are connected to input bit I:001/01 and the NO contacts of the start PB are connected to input I:001/00. So, if the Start PB is depressed the output O:003/01 is latched on. When the start bit (I:001/01) goes false, the output instruction for the pump starter will remain ON until the stop bit I:001/00 is depressed to unlatch the output. Note that the output unlatch instruction has the same address as the latched bit.

The LD program in Figure 7-5 is a simpler method to produce a start/stop function then the standard Start/Stop LD program shown in Figure 7-3. However, this latch and unlatch pump start/stop program must be used with caution because if power is lost to the PLC the pump will turn off but when power is restored the pump will automatically restart. The control system programmer must make sure that this is a safe procedure for a given application.

## One Shot (ONS)

The one shot is an input instruction that goes true for one PLC program scan, if there is a false to true transitions in the conditions preceding it in the rung. The one shot instruction is generally used to start operations that are triggered by momentary pushbutton action, such as obtaining values from thumbwheel switches or freezing rapidly displayed LED data. In the Allen-Bradley series 5 PLCs, the bit address must be either a binary file (B3) or an integer file (N7) bit address. A typical example is shown in Figure 7-6. In this application, when the

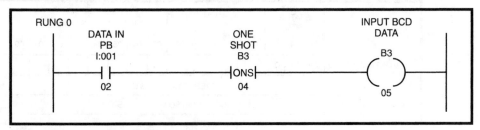

**Figure 7-6. One shot (ONS) application.**

data in pushbutton (PB) is depressed it sets input bit (I:001/02) to 1 and the ONS bit (B3/04) conditions the rung so that the output (B3/05) turns on for one scan. The output turns off for successive scans until the input goes from false to true again.

## Timer and Counter Instructions

Timers and counters are output instructions that provide the same functions as would hardware timers and counters. They are used to activate or deactivate a device after an expired interval or count. The timer and counter instructions are generally considered internal outputs. Like the relay-type instructions, timer and counter instructions are fundamental to the ladder logic instruction set.

The operations of timers and counters are quite similar in that they are both counters. A timer counts the number of times that a fixed interval of time of either 0.01 seconds or 1.0 sec elapses in A-B PLC5s. To time an interval of 3 seconds, a timer counts three 1-second intervals or three hundred 0.01 second time intervals. A counter simply counts the occurrence of an event. The timer and counter instructions require an accumulator (ACC) register (word location) to store the elapsed count and a preset (PR) register to store a preset value. The preset value will determine the number of event occurrences or time-based intervals that are to be counted. When the accumulated value equals the preset value, a status bit is set on and can be used to turn on an output device.

Since timer or counter instructions store an accumulated and preset value, three words of memory are required for these instructions in A-B PLC5s timers as shown in Figure 7-7.

The leftmost three bits in the first timer word (bits 13, 14, and 15) are used as status bits. Bit 15 is the timer enable (EN) bit and it is set when the rung goes true. Bit 14 is the timer timing bit (TT) and it is set when the timer rung goes true, and it indicates that a timing operation is in progress. Bit 13 is the timer done (DN) bit and it is true when the accumulated value is greater than or equal to the preset value.

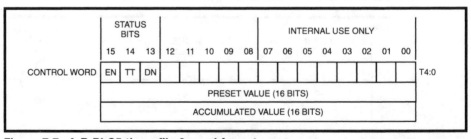

**Figure 7-7. A-B PLC5 timer file 3 word format.**

## Timer On Delay (TON)

The timer on delay (TON) output instruction is programmed to provide time delayed action or to measure the duration for which some event is occurring. If any rung path connected to the input side of the timer has logic continuity as shown in Figure 7-8, the timer begins counting time-based intervals and counts until the accumulated (ACCUM) time equals the preset value as long as the rung conditions remain true. When the accumulated time equals the preset time, a timer done (DN) bit in the timer word is set to 1. Whenever the rung logic conditions for the TON instruction go false, the accumulated value is reset to all zeros.

In the example application shown in Figure 7-8, when the pump start switch is ON, bit I:000/01 is set to 1 and the timer (T4:0) begins to count time-based intervals. As long as the switch remains closed or ON, the timer increments its accumulated value word for each time interval. When the accumulated value equals the preset value of 5 seconds, the timer stops incrementing its accumulated value and sets the timer done (DN) bit to on. This done bit (T4:0/DN) is then used in rung 1 to energize the run pump output bit (O:001/01).

## Timer Off Delay (TOF)

The timer off delay output instruction provides another form of timer action. If logic continuity is lost, the timer begins counting time-based intervals until the accumulated time equals the programmed preset value. When the accumulated time equals the preset time, the timing is complete, and the timer done bit (bit 13) is set to zero. The timer done (DN) bit can be used throughout the program as a NO or NC contact. If logic continuity is gained before the timer is timed out, the

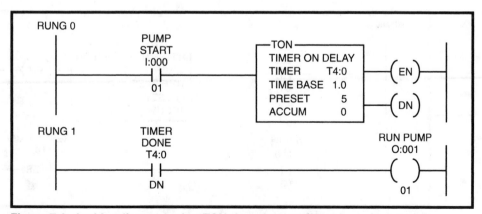

Figure 7-8. Ladder diagram using TON timer instruction.

accumulator word is set to zero. An example program for a TOF instruction with a preset value of 5 seconds is shown in Figure 7-9.

In rung 0, if the input switch (I:000/01) is open for 5 seconds the timer will count up to 5. When the preset value is equal to the accumulated value of 5, the timer done bit (T4:1/DN) will be set to 0 and the delay complete output bit O:001/01 in rung 1 will be turned on.

## Retentive Timer On (RTO)

The retentive timer on output instruction is used if it is necessary for the timer accumulated value to be retained, even if logic continuity or power is lost. If the timer rung path has logic continuity, the timer begins counting time-based intervals until the accumulated time equals the preset value. The accumulator register retains the accumulated value, even if logic continuity is lost before the timer is timed out or if power is lost, as shown in Figure 7-10. When the accumulated time equals the preset time, the output is energized, and the timed out contact associated with the output is turned ON. The timer contacts can

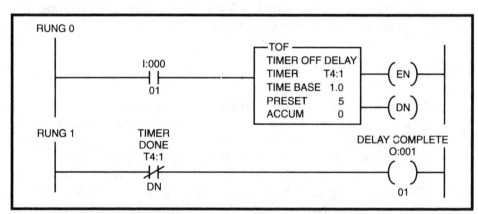

Figure 7-9. Ladder diagram using a TOF timer instruction.

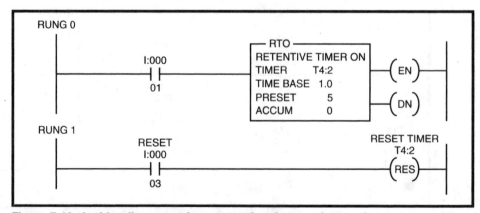

Figure 7-10. Ladder diagram using a retentive timer on instruction.

be used throughout the program as a NO or NC contact. The retentive timer accumulator value must be reset to zero by a reset (RES) instruction as shown in Figure 7-10. This instruction resets timers and counters, as well as control blocks.

A typical application for timers is to produce alternating pulses for a flashing alarm light as illustrated in the following example problem.

---

**Example 7-3**

**Problem:** Design a simple timing circuit that can be used to generate an alternating signal to produce a flashing alarm light connected to output at bit location O:001/003.

**Solution:** The ladder logic program consists of two TON instructions driving an alarm light as shown in Figure 7-11.

---

## Up Counter (CTU)

The up counter output instruction will increment by one each time the counted event occurs. A control application of a counter is to turn a device ON or OFF after reaching a certain count. An accounting application of a counter is to keep track of the number of filled cans that pass a certain point. The up counter increments its accumulated

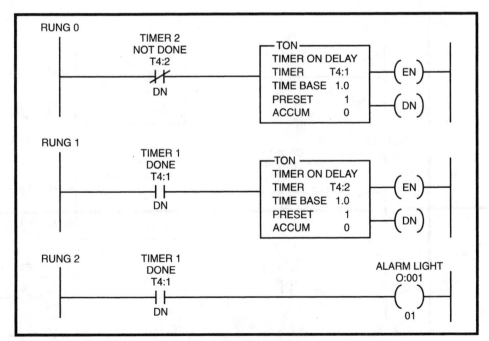

**Figure 7-11. Ladder digram for flashing alarm light.**

value each time the up-count logic input makes an OFF-to-ON transition. Since only the false-to-true transition causes a count to be accepted, the rung condition must go from true to false and back to true before the next count is registered.

When the accumulated value reaches the preset (PR) value, the output is turned ON, and the COUNT DONE bit is to one. Unlike a timer instruction, the counter instruction continues to increment its accumulated value after the preset value has been reached. If the accumulated value goes above the maximum range of the counter, an overflow (OV) bit will be set. This overflow bit can be used to cascade counters for counter applications greater than the maximum value of the counter.

## Down Counter (CTD)

The down counter output instruction will count down by one each time a certain event occurs. Each time the down-count event occurs, the accumulated value is decremented. In normal use, the down counter is used in conjunction with the up counter to form an up/down counter.

Figure 7-12 shows a typical example of an A-B PLC5 Up/Down Counter instruction. Note that the same word address, C5:0, is used for both counters.

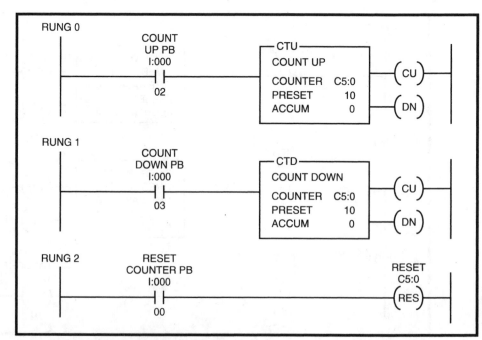

**Figure 7-12. Ladder diagram for up/down counter.**

---

**EXAMPLE 7-4**

**Problem:** Design an up/down counter circuit to count the number of parts produced on an assembly line. Assume input I:000/12 of the PLC is activated by each part leaving the final assembly line; input I:000/13 is activated by a part being rejected in final test, and input I:000/00 is energized at the end of a production run.

**Solution:** The ladder program to calculate the number of parts produced is given in Figure 7-13.

---

# Data Transfer Operations

Data transfer instructions involve the transfer of the contents from one register to another. Data transfer instructions can address any location in the memory data table, with the exception of areas restricted to user application. Pre-stored values can be automatically retrieved and placed in any new location. That location may be the preset register for a timer or counter or even an output register that controls a seven-segment display.

The Allen-Bradley PLC5 programming system uses three data bit and word transfer instructions: bit distribute (BTD), move (MOV), and masked move (MVM). These data transfer instructions are typical for most PLC manufacturers.

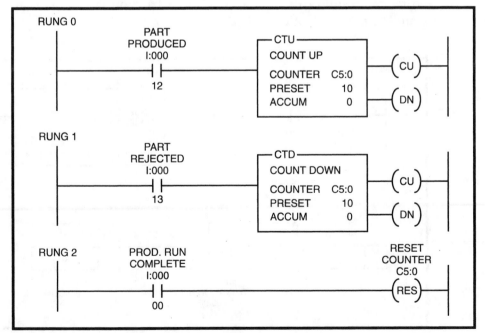

**Figure 7-13. Ladder diagram for production part count.**

## Bit Distribute (BTD)

The BTD instruction is an output instruction that moves up to 16 bits of data within or between words and the source remains unchanged. Figure 7-14 shows a BTD example of moving bits within a word. The instruction writes over the destination with the specified bits. If the length of the bit field extends beyond the destination word, the processor does not save the overflow bits. These overflow bits are lost because they do not wrap into the next word.

On each processor program scan, when the rung that contains the BTD instruction is true, the processor moves the bit field from the source word to the destination word. To move the data within a word, the programmer selects the same word address for both the source and the destination as shown in Figure 7-14. In this example, four bits are moved from the left hand side (bits 00 through 03) of word N70:00 to the middle of the word (bits 08 through 11).

## MOVE (MOV) and MASKED MOVE (MVM)

The MOV instruction is an output instruction that copies the source address to the destination address. As long as the rung remains true, the instruction moves the contents each scan. The source is a program constant or data address from which the instruction reads an image of the value. The destination is the data address to which the instruction writes the result of the operation. The instruction writes over any data stored at the destination.

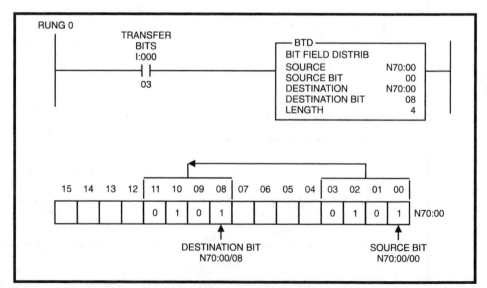

**Figure 7-14. Example of BTD instruction moving bits within a word.**

The MVM instruction is an output instruction that copies the source to a destination and allows portions of the data to be masked. As long as the rung remains true, the instruction moves data each scan.

The MVM instruction can be used to copy I/O image table, binary, or integer values. For example, the MVM instruction can be used to extract bit data such as status or control bits from an address that contains bit and word data. The source is a program constant or data address from which the instruction reads an image of the value and the source remains unchanged.

The mask can be an address or hexadecimal value that specifies which bits to pass or block. The programmer must set the mask bits to 1 to move the data. The moved data overwrites the destination data. The destination is the data address to which the instruction writes the result of the operation. The instruction writes over any data stored at the destination. Figure 7-15 shows an example of both move and masked move instructions.

## Arithmetic Operations

The arithmetic operations include the four basic operations of addition, subtraction, multiplication, and division. These instructions use the contents of two word locations and perform the desired function.

The arithmetic instructions are programmed in the output portion of the ladder and use either one or two data words to store the result. The add and subtract instructions use one word. Multiply and divide need two words for the computed result.

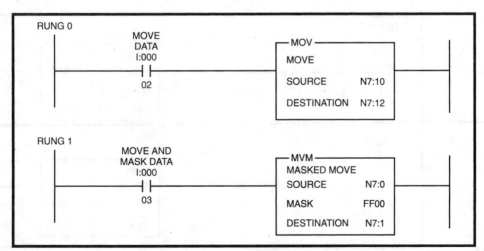

**Figure 7-15. Typical example of MOV and MVM instructions.**

## Addition (ADD)

The ADD instruction performs the addition of two values stored in two different memory locations, source A and source B, and the result is placed in the destination. Source A and source B can be either values or addresses that contain values. In the example shown in Figure 7-16, if the input I:000/01 is set, the instruction adds the value in N7:0 to the value of N7:1 and stores the result in address N7:2.

## Subtraction (SUB)

The SUB instruction performs the subtraction operation of two numbers location in source A and B. As in addition, if a condition to enable the subtraction is set to 1, the result is placed in a destination. In Figure 7-16, looking at rung 1, if the input bit at address I:000/02 is set to 1, the number in address N7:4 is subtracted from the number in address N7:3 and the result is placed in register N7:5.

## Multiplication (MUL)

The MUL instruction is used to multiply one value (Source A) by another value (Source B) and to place the result in the destination. Source A and Source B can be values or addresses. In the typical MUL instruction example of Figure 7-17, if input bit I:000/03 in rung 0 is true, the processor will multiply the value in N7:3 by the value in N7:4 and store the result in N7:20.

## Divide (DIV)

The DIV instruction is used to divide one value (Source A) by another value (Source B) and place the result in the destination. Source A and

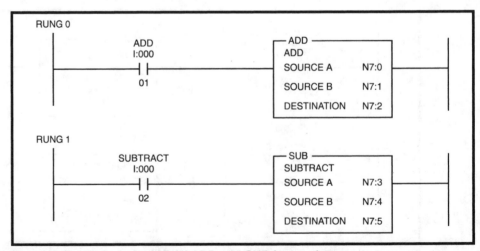

**Figure 7-16. Typical example of ADD and SUB instructions.**

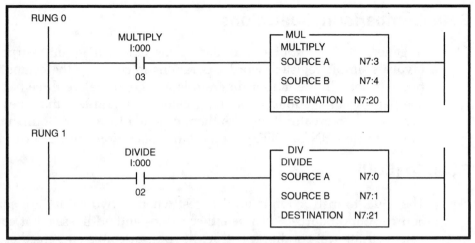

**Figure 7-17.  Typical examples of MUL and DIV instructions.**

Source B can be values or addresses. In the typical DIV instruction example of Figure 7-17, if input bit I:000/02 in rung 1 is true, the processor will divide the value in N7:0 by the value in N7:1 and store the result in N7:21.

A typical math application in programmable controller systems is to scale process analog input signals, as illustrated in the next example.

---

**EXAMPLE 7-5**

**Problem:**  An analog 4 to 20 ma signal from a pressure transmitter has a range of 0 to 20 inches. This signal is converted to 0 to 100% in the analog input module and stored in F8:20. Write a ladder logic program to convert the signal back to engineering units and store the result in location F8:25.

**Solution:**  The first step is to calculate the scale factor as follows:

$$\text{Scale factor} = \frac{\text{pressure range}}{\text{PLC input range}}$$

$$= \frac{(0 - 20) \text{ inches}}{(0-100)\%}$$

$$= 0.2 \text{ inches}/\%$$

We now use this scale factor in the ladder diagram as shown in Figure 7-18.

---

## Data Comparison Operations

In general, the manipulation of data using ladder diagram instructions involves simple register (word) operations to compare the contents of two registers. In the ladder diagram language, there are three basic data comparison instructions: equal to, greater than, and less than. Based on the result of a greater than, less than, or equal to comparison, an output can be turned ON or OFF, or some other operation can be performed.

### Equal To (EQU)

The equal to instruction is used to test whether two values are equal. Source A and Source B can be either values and addresses that contain values. If the test condition is true, the output coil is energized as shown in Figure 7-19.

### Less Than (LES)

Similar to the equal to instruction, the LES instruction tests the contents of the value of one location (Source A) to see if it is less than the value stored in a second location (Source B). If the test condition is true, the output coil is energized as shown in Figure 7-20.

### Greater Than (GRT)

The greater than instruction operates like the less than operation, with the exception that the test is performed for a greater than condition. If

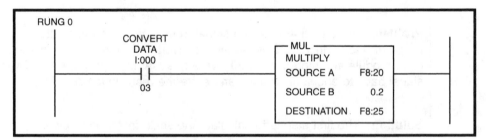

**Figure 7-18. Process data scale convertion.**

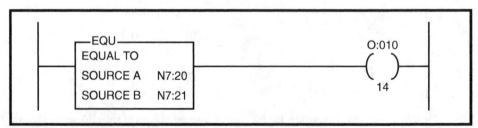

**Figure 7-19. Typical equal to instruction.**

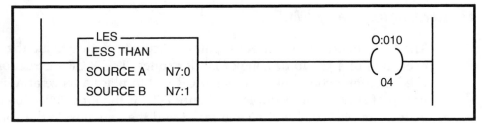

Figure 7-20. A typical less than instruction.

the test condition is true, the output coil is energized as shown in Figure 7-21.

---

**EXAMPLE 7-6**

**Problem:** Write a ladder logic program to energize output coil O:001/03, if the data in word N7:0 is less than the data in word N7:1.

**Solution:** The required ladder logic program is shown in Figure 7-22.

---

# Program Control Instructions

The program control functions are used to perform a series of conditional and unconditional jump and return instructions. These instructions allow the program to execute only certain sections of the control logic if a fixed set of logic conditions are met. The following instructions are a representative selection of some of the program control instructions available in most controllers.

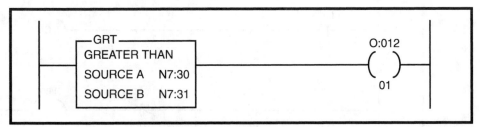

Figure 7-21. A typical greater than instruction.

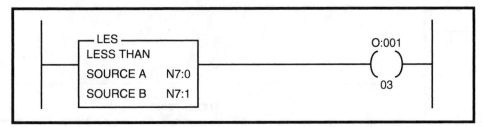

Figure 7-22. Less than ladder logic rung.

## Master Control Relay (MCR)

The MCR instruction is used in pairs to activate or deactivate the execution of a group or zone of ladder rungs. The MCR instruction is used in conjunction with another MCR coil to place a fence around the group of rungs. For example, in Figure 7-23, if the input I:000/03 is active, the MCR coil will be energized and the logic inside the zone will be executed by the controller. If the MCR coil is turned OFF, all outputs inside the zone will be deenergized.

The rungs within an MCR zone are still scanned, but the PLC scan time is reduced due to the false state of the nonretentive outputs. Nonretentive outputs are reset when their rung goes false.

## Jump (JMP) and Label (LBL)

The JMP and LBL instructions are used in pairs to skip portions of the ladder logic program. The jump instruction allows the normal sequential execution to be altered so that the CPU will jump to a new position in the ladder program. If the jump rung logic is true, the jump coil (JMP) instructs the CPU to jump to and execute the rung labeled with the same reference address as the jump coil. This allows the program to execute rungs out of the normal sequential flow of a standard ladder program.

The label is to identify that ladder rung that is the destination of a jump instruction. The label reference must match that of the jump

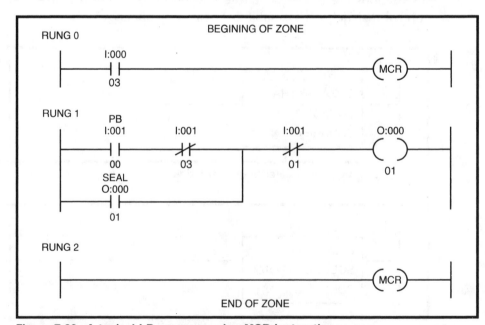

Figure 7-23. A typical LD program using MCR instructions.

instruction with which it is used. The label instruction does not
contribute to logic continuity, and it is always logically true. It is placed
as the first logic condition in the rung. A label instruction referenced by
a unique address can be defined only once in a program.

## Jump to Subroutine (JSR), Subroutine (SBR), and Return (RET)

The JSR, SBR, and RET instructions in A-B PLC5 direct the processor to
go to a separate subroutine file within the ladder logic program, scan
that subroutine file once, and return to the point of departure.

The JSR instruction directs the processor to the specified subroutine file
and, if required, defines the data passed to and received from the
subroutine. The optional SBR instruction is the instruction that stores
incoming data. The SBR instruction is only used if the ladder diagram
requires that data is passed to and from the subroutine. The RET
instruction ends the subroutine and, if required, stores data to be
returned to the JSR instruction in the main program. If the SBR
instruction is used, it must be the first instruction on the first rung in
the program file that contains the subroutine.

Subroutines are used to store recurring sections of ladder logic that can
be accessed from different parts of the main ladder logic program. A
routine saves memory space because it is program only once.

# Instruction List Language (IL)

IL is a low level IEC 1131-3 standard PLC language. It is highly
effective for smaller applications or for optimizing parts of an
application. Instructions always relate to the current result or IL
register. The operator indicates the operation that must be made
between the current value and the operand. The result of the operation
is stored again in the current result.

## IL Language Syntax

The IL program is simply a list of instructions. Each instruction must
begin on a new line, and must contain an operator, completed with
optional modifiers and, if necessary, for the specific operation, one or
more operands, separated with commas (","). A label followed by a
colon (":") may precede the instruction. If a comment is attached to the
instruction, it must be the last component of the line. Comments always
begin with "(*" and end with "*)". Empty lines may be entered
between instructions. Comments may be placed on empty lines. An
example of an instruction list program follows:

| Label  | Operator | Operand | Comments |
|--------|----------|---------|----------|
| Start: | LD       | IX1     | (* load input IX1, start pushbutton*) |
|        | ANDN     | MX5     | (* AND with NOT of MX5) |
|        | ST       | QX2     | (* store output QX2 to start motor*) |

Labels are used as operands for some operations such as jumps, but they are generally used to document a program for clarity. Naming labels must conform to the following rules: labels cannot exceed 16 characters; the first character must be a letter; characters following the first letters must be letters, digits, or the "_" character. The same name cannot be used for more than one label in the same IL program. A label can have the same name as a variable.

There are three modifiers used with IL operators. The first is the capital letter "N"; it is used to produce a Boolean negation of an operand such as an input bit. For example, the instruction ORN IX12 is interpreted as a NOT OR operation on the input bit IX12. The next modifier is the parenthesis character "(" and it indicates that the evaluation of the instruction must be delayed until the closing parenthesis ")" modifier is encountered. The final modifier is the capital letter "C" and this modifier indicates that the attached instruction must be executed only if the current result has the Boolean value of True (different than 0 for non-Boolean values). The "C" modifier can be combined with the "N" modifier to indicate that the instruction must be executed only if the current result has the Boolean value FALSE (or 0 for non-Boolean values). Table 7-1 summarizes the standard operators and modifiers used in the IL language.

The basic IL operators listed in Table 7-1 can be used to perform basic logic control program. The following IL program demonstrates several important operators and their modifiers.

| Example Instruction List Program: | | |
|---|---|---|
| LD   | IX1 | (*current result: = TRUE*) |
| AND  | IX2 | (*current result: = IX1 AND IX2*) |
| ANDN | IX3 | (*current result: = IX1 AND IX2 and NOT (IX3)*) |
| ST   | QX0 | (*QX0: = current result*) |

## PLC Control Program Documentation

An important part of PLC programming is the proper and complete documentation of the control program. Most PLC manufacturers provide a means to print out a hard copy of the control program stored in the PLC's memory. Whether stored in ladder diagram form or some

| Operators | Modifiers | Operand | Description |
|---|---|---|---|
| LD | N | Variable, constant | Loads operand |
| ST | N | Variable | Stores current result |
| S |  | Boolean variable | Sets to TRUE |
| R |  | Boolean variable | Resets to FALSE |
| AND (&) | N( | Boolean variable | Boolean AND |
| OR | N( | Boolean variable | Boolean OR |
| XOR | N( | Boolean variable | Exclusive OR |
| ADD | ( | Variable, constant | Addition |
| SUB | ( | Variable, constant | Subtraction |
| MUL | ( | Variable, constant | Multiplication |
| DIV | ( | Variable, constant | Division |
| GT | ( | Variable, constant | Greater than |
| GE | ( | Variable, constant | Greater than or equal to |
| EQ | ( | Variable, constant | Equal to |
| LE | ( | Variable, constant | Less than or equal to |
| LT | ( | Variable, constant | Less than |
| JMP | C, N | Label | Jump to label |
| Ret |  |  | Return from sub-program |
| ) |  |  | Delayed Operation |

**Table 7-1. IL Language Operators.**

other language, the hardcopy will be an exact replica of the control program stored in memory.

This hardcopy printout will show each programmed instruction with the associated address of each input and output. However, the information on the function or purpose of each field device or internal control bit or instruction is not readily apparent; additional documentation is generally required. Most PLC manufacturers provide software documentation programs that allow the programming device, generally a personal computer, to enter labels or mnemonic nomenclature for each program element or instruction. The example ladder diagram programs in this chapter illustrate documented elements or instructions such as input, output, timer, counter, etc. in the ladder rungs. Most PLC programming software packages also allow for general comments on each rung of logic as shown in Figure 7-24.

The PLC controller will always have the latest revision of the control program stored in memory. So before testing a program on-line, the user should printout the latest revision of the program. During start-up and testing, frequent changes are made to the control program, which should be immediately documented with both rung and instruction comments. It is also good practice to obtain the latest hardcopy of the program at the earliest convenience.

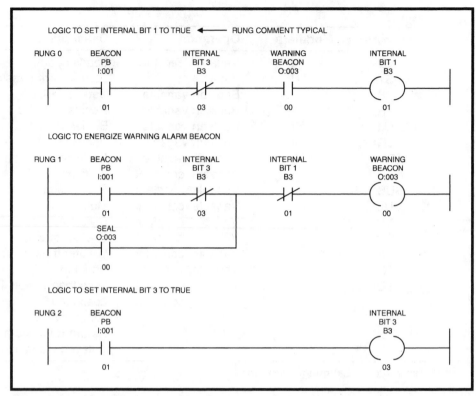

**Figure 7-24. Alarm Beacon application program with rung comments.**

## EXERCISES

7.1   Discuss the various PLC programming standard languages.

7.2   Write a ladder diagram program to manually control an electric motor. Assume that a start pushbutton is wired to an input module at I:000/00 and a stop pushbutton is connected to the same input module at address I:000/01. Also assume that the start PB is normally open and the stop pushbutton is normally closed. An output module point will drive a motor control relay at address O:000/03, and there is an auxiliary motor start contact (NO) connected to PLC input I:000/02.

7.3   Revise the pump control ladder program shown in Figure 7-3 using latch and unlatch output instructions instead of the standard output coil shown.

7.4   Write a ladder logic program to control a process pump under the following conditions:

1. Turn on the pump (output bit O:001/00) five seconds after the inlet valve (input bit I:000/02) and the outlet valve (input bit I:000/03) to the pump have been opened.
2. Turn off the pump if either the inlet or the outlet valve to the pump is closed.

7.5    Design a ladder program using timer instructions to flash four process alarm lights on a control panel at a rate of 0.3 second. The alarm lights are driven by PLC outputs at O:010/00, O:010/01, O:010/02, and O:010/03. Connect a panel-mounted pushbutton to input point I:000/00 to acknowledge the alarms. Assume the alarm lights will remain ON after the acknowledge PB is depressed and will turn OFF only if the alarm input has been cleared. Complete a control panel layout and I/O wiring diagrams for the system.

7.6    A temperature transmitter connected to an analog input module has a range of 100°C to 500°C. This signal is converted from 0 to 100% in the programmable controller. Design a ladder logic program to convert the signal back to engineering units.

7.7    Design a ladder program to control three heater elements in a process fluid storage tank used to heat fluid to 400°C. Assume that the heater contactors are energized at 1-second intervals to produce a smoother heat curve and better energy conversion. Use ON/OFF control to maintain the temperature at the set point.

7.8    Write a ladder logic program to transfer some data in word N7:0 to an output display at word N7:10, if the data in word N7:0 is greater than the data in word N7:1 and less than the data in location N7:2.

7.9    Write an IL language program to start and stop an electric pump. Assume the following: a normally open start pushbutton connected to input bit, IX1, a normally closed stop pushbutton is connected to input, IX2, a discrete output drives a pump starter relay at address QX1, and an auxiliary motor start contact (NO) connected to input point IX3.

7.10   Explain the purposes and benefits of PLC control program documentation.

## BIBLIOGRAPHY

1.  Jones, C. T., and Bryan, L. A., *Programmable Controllers: Concepts and Applications*, International Programmable Controls, Inc., First Edition, 1983.

2.  *Programmable Controller Fundamentals*, Allen-Bradley Company, 1985.

3.  *PLC-5 Programmable Controller: Instruction Set Reference*, Rockwell Software, Inc., 1996.

4.  Hughes, T. A., *Basics of Measurement and Control*, Instrument Society of America, Second Edition 1995.

5.  Gilbert, R. A., and Llewellyn, J. A., *Programmable Controllers: Practices and Concepts*, Industrial Training Corporation, 1985.

6.  Reis, R. A., and Webb, J. W., *Programmable Controllers: Principles and Applications*, Prentice-Hall, Inc., Third Edition, 1995.

7.  Bryan, E. A., and Bryan, L. A., *Programmable Controllers: Concepts and Applications*, Industrial Text Co., First Edition, 1988.

# 8

# Advanced PLC Programming

## Introduction

The advanced programmable controller languages and instructions are required to perform more powerful functions beyond simple ON/OFF control, timing, counting, and data manipulation. These advanced languages are used for analog control, data file operations, sequencer operations, data reporting, complex logic functions, and other functions that are not possible with the basic programming languages. The four advanced languages to be discussed are ladder diagram (LD) with advanced function block instructions, structured text (ST), function block diagrams (FBD), and sequential function charts (SFC). The most common advanced language used is LD with advanced function block instructions, so it will be discussed first and in the greatest detail.

## Advanced Function Block Instructions

Function block instructions allow the user to program more complex PLC control functions. The most common advanced instructions—file, shift register, sequence, and block transfer—will be discussed.

### File Instructions

A file is a group of consecutive data table words used to store PLC information. A file instruction is used to perform arithmetic, logical, search, copy, and compare operations. We will only discuss the Allen-Bradley PLC5 file instructions—file arithmetic and logical (FAL), file search and compare (FSC), file copy (COP), and file fill (FLL)—but most PLC manufacturers have similar instructions.

The structure of a typical file instruction is shown in Figure 8-1. This figure illustrates an FAL with its parameters of control, length, position, mode, destination, and expression.

The *control* is the address of the control structure in a control type (R) file. The processor uses this information to run the instruction. The *length* is the number of words (1 to 1000) in the data block on which the file instruction operates. The *position* is the current block within the data block that the processor is accessing. The *mode* is the number of file words operated on each time the rung is scanned in the program. There are three modes: All, numerical, and Incremental. In the All mode, the entire file is operated on before continuing on to the next rung of the program. The numerical mode distributes the file operation over a number of program scans. The Incremental mode manipulates one word of the file each time the rung goes from false to true. The *destination* is the address where the processor stores the result of the operation. The instruction converts to the data type specified by the destination address. The *expression* contains addresses, program constants, and operators that specify the source of data and the operations to be performed.

The output coils to the right of the file instruction are the enable (EN), done (DN), error (ER) bits. These bits have the same word address as the instruction control. The address of these status bits is automatically set by the processor when the control address is entered by the programmer. The enable (EN) bit is set by a false-to-true rung transition and indicates the instruction is enabled. In the incremental mode, the EN bit follows the rung condition. In the numerical and All modes, the EN bit remains set until the instruction completes its operation, regardless of the rung condition. The enable bit is reset when the rung goes false and the instruction completes its operation.

The done (DN) bit is set after the instruction has operated on the last set of words. In the numerical mode if the instruction is false at completion, it resets the done bit one program scan after the operation

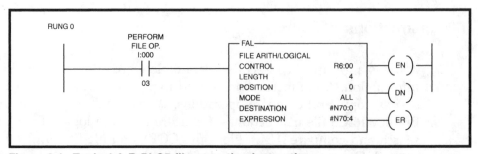

**Figure 8-1. Typical A-B PLC5 file operation instruction.**

is complete. If the instruction is true at completion, the done bit is reset when the instruction goes false.

The error (ER) bit is set when the operation generates an overflow. The instruction stops until the ladder program resets the error bit. When the processor detects an error, the position value stores the number of the word that faulted.

## File Arithmetic and Logic (FAL)

The FAL instruction performs copy, arithmetic, logic, and function operations on the data stored in files. The FAL instruction is an output instruction that perform the operations defined by the source addresses and the operators listed by the programmer in the expression field. The instruction writes the results into a destination address. The FAL instruction automatically converts the data type at the source addresses to the data type that is specified in the destination address.

The FAL instruction performs operations such as zero a file, copy data from one file to another, make arithmetic or logic computations on data stored in files, and unload a file of error codes one at a time for display. Table 8-1 lists the operations performed by the FAL instruction.

| Type | Operator | Description | Example Operation |
|------|----------|-------------|-------------------|
| Copy | none | Copy from A to B | Enter source address in expression |
| Clear | none | Set a value to 0 | enter 0 for expression |
| Arithmetic | + | Add | 2+2 |
| | − | Subtract | 8−5 |
| | * | Multiply | 3*6 |
| | \| | Divide | 12\|4 |
| | − | Negate | −N7:0 |
| | SQR | Square root | SQR N7:1 |
| | ** | Exponential | 10**2 |
| Bitwise | AND | Bitwise AND | D9:3 AND D10:4 |
| | OR | Bitwise OR | D9:4 OR D9:5 |
| | XOR | Bitwise exclusive or | D9:6 XOR D9:7 |
| | NOT | Bitwise complement | NOT D10:11 |
| Conversion | FRD | Convert from BCD to binary | FRD D10:0 |
| | TOD | Convert from binary to BCD | TOD N7:1 |

**Table 8-1. FAL Operators.**

To illustrate the operation of a FAL instruction, we will perform a FAL copy operation as shown in Figure 8-2.

In this example, when the rung goes true (bit I:000/02 set to 1), the processor reads four words of integer file N71 starting at word 3, and writes the image to integer file N70 starting at word 0. It writes over any data in the destination file.

---

### EXAMPLE 8-1

**Problem:** Write a LD program to copy the data in integer file N30, words 5, 6 and 7 to file N31 starting at word 2 if input bit I:000/03 is true.

**Solution:** The ladder diagram program to copy the data is shown in Figure 8-3.

---

## File Search and Compare (FSC)

The FSC instruction is an output instruction that compares values in source files, word by word, for the logical operation that is specified in

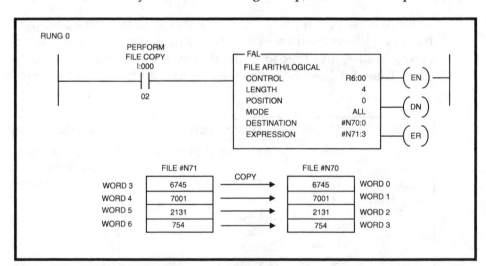

**Figure 8-2. A-B PLC5 file-to-file copy example.**

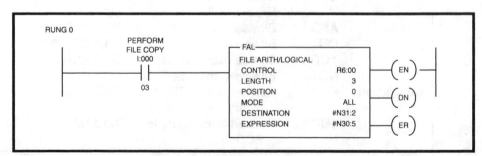

**Figure 8-3. File-to-file copy LD for Example 8-1.**

the expression. When the processor finds the specified comparison is true, it sets the found bit (FD), and records the position where the true comparison was found. The inhibit bit (IN) is set to prevent any further searching of the files.

This instruction is used to perform operations such as set high and low process alarms for multiple analog inputs and compare batch variables against a reference file before starting a batch operation.

The FSC instruction performs the comparisons listed in Table 8-2 on file data according to equation listed in the expression portion of the instruction.

The processor compares files of different data types by internally converting data into its binary equivalent before performing the comparison.

In file search, when the rung condition is true, the desired comparison is performed on data addressed in the expression. Words are compared in ascending order, starting at the beginning. The rate is determined by the mode of operation that is specified in the FSC instruction. The done (DN) bit is set after the processor has compared the last pair. If the rung is true at completion, the done bit is turned off when the rung is no longer true. In the numerical mode, however, if the rung is not true at completion, the DN bit stays on one program scan after the operation is complete.

To illustrate the operation of a FSC instruction, we will perform an FSC "search not equal" operation as shown in Figure 8-4. When bit I:000/03 goes to true, the processor performs the "not-equal-to" comparison between words, starting at B4:0 and B5:0. The number of words compared per program scan is 10 in this example because the mode is set to 10.

When the processor finds that corresponding source words are not equal (words B4:4 and B5:4), the processsor stops the search and sets the found (FD) and inhibit (IN) bits. To continue the search comparison the ladder logic program must turn off the inhibit bit.

| Comparison | Example Expression |
|---|---|
| Search equal | #N50:0 = #N51:0 |
| Search not equal | #N51:0<>#N53:10 |
| Search less than | #N51:0<#N53:10 |
| Search less than or equal | #N51:0< = #N53:10 |
| Search greater than | #N51:0>#N53:10 |
| Search greater than or equal | #F61:0> = #N63:10 |

Table 8-2. FSC Comparisons.

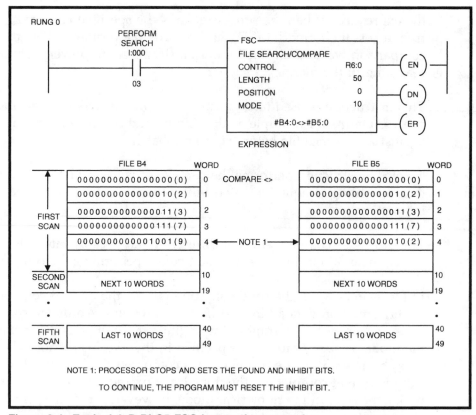

**Figure 8-4. Typical A-B PLC5 FSC instruction example.**

---

**EXAMPLE 8-2**

**Problem:** Write an LD program to search the data in integer file N40, words 0 through 99 and compare for an equal condition to data in file N50 starting at word 0, if input bit I:000/03 is true.

**Solution:** The ladder diagram program to search and compare the data files is shown in Figure 8-5.

---

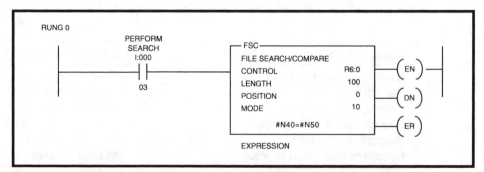

**Figure 8-5. FSC ladder diagram to perform search and compare equal.**

## File Copy (COP)

The file copy instruction is an output instruction that copies the values in the source file into the destination file. The source file remains unchanged. The COP instruction does not use status bits.

The COP instruction will not write over file boundaries, so any overflow data will be lost. Also, no data conversion occurs, so the source and destination files should use the same data type.

If the destination is in a file of words (such as an integer file), the programmer will specify the length in words. If the destination is in a file of structures, such as a counter file, the programmer must specify the length in structures. For example, if the source word is an integer file and the destination is a counter file, the programmer would specify a length of 5, the source word is copied 15 times to fill 5 counter structures with 15 words.

An example LD program using a COP instruction is illustrated in Figure 8-6. In this example, if input bit I:000/03 is true, the processor will copy the first ten words starting at file N50:0 into the first ten words of file N60:0.

## File Fill (FLL)

The file fill instruction is an output instruction that fills the words of a file with a source value. The source file remains unchanged. Like the COP instruction, the FLL instruction does not use status bits.

The FLL instruction will not write over file boundaries, so any overflow data will be lost. Also, no data conversion occurs, so the source and destination files should use the same data type.

If the destination is in a file of words, such as an integer file, the programmer will specify the length in words. If the destination is in a file of structures, such as a counter file, the programmer must specify

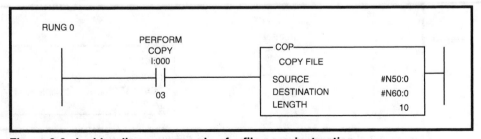

**Figure 8-6. Ladder diagram example of a file copy instruction.**

the length in structures. For example, if the source word is an integer file and the destination is a counter file, the programmer would specify a length of 5, the source word is copied 15 times to fill 5 counter structures with 15 words.

An example LD program that uses an FLL instruction is illustrated in Figure 8-7. In this example, if input bit I:000/03 is true, the processor will copy the first ten words starting at file N50:0 into the first ten words of file N60:0.

## Shift Register Instructions

Shift register instructions are used to track the movement or flow of parts and information in industrial applications. We will discuss four A-B PLC5 shift register instructions in common use: bit shift left (BSL), bit shift right (BSR), first in-first out load (FFL), first in-first out unload (FFU).

The bit shift instructions, BSR and BSL, are used to load bits into, shift bits through, and unload bits from a bit array one bit at a time, such as for tracking bottles through a bottling line where each bit represents a bottle. The load and unload shift instructions, FFL and FFU, are used to load and unload values in the same order, such as for tracking parts through an assembly line where parts are represented by values that have a part number and assembly code.

### Bit Shift Instructions

Bit shift instructions shift all bits within the specified address one bit position with each false-to-true ladder rung transition. There are two shift instructions: bit shift left (BSL) and bit shift right (BSR)

The structure of a bit shift left instruction is shown in Figure 8-8. This figure illustrates a BSL instruction with its parameters of file, control, bit address, and length.

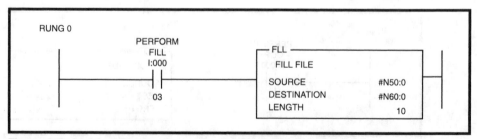

**Figure 8-7. Ladder diagram example of a file fill instruction.**

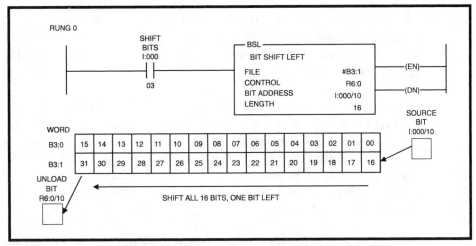

**Figure 8-8. An example of bit shift left (BSL) instruction.**

The *file* is the address of the bit array that will be manipulated. The array must start at a 16-bit word boundary. For example, use bit 0 of word number 1, 2, 3, etc. The array can end at any bit number up to 15,999. However, the remaining bits in the last word of the array cannot be used because the instruction invalidates them.

The *control* is the address of the control structure (48 bits—three 16-bit words) in a control type (R) file that stores the instruction's status bits, the size of the array (number of bits), and the bit pointer. The processor uses this information to run the instruction.

The *bit address* is the address of the source bit. The instruction inserts the value (0 or 1) of this bit in either the first (lowest) bit position (for the BSL instruction) or the last (highest) bit position (for the BSR instruction) in the array.

The *length* is the decimal number of bits to be shifted. In A-B PLC5s the bits in I/O files are numbered in octal 00 to 07 and 10 to 17, but all other files are numbered in decimal 0 to 15.

The output coils to the right of the file instruction are the enable (EN) and done (DN) bits. These bits have the same word address as the instruction control element. The address of these status bits is automatically set by the processor when the control address is entered by the programmer. The enable (EN) bit is set by a false-to-true rung transition and indicates that the instruction is enabled. The done (DN) bit is set to indicate that the bit array shifted one bit position.

The control element also contains two other status bits: error (ER) at bit 11 and unload (UL) at bit 11. The error (ER) bit is set to indicate that the instruction detected an error, such as if you entered a negative file length. The unload (UL) bit is the instruction's output. The UL bit stores the status of the bit removed from the array each time the instruction is enabled.

In the example in Figure 8-8, when the rung containing the BSL instruction goes from false to true, the processor sets the EN bit. Then the processor shifts 16 bits (length = 16) in bit file B3, starting with bit 16, to the left (higher bit number) one bit position. The last bit shifts out at bit position 31 into the control register UL bit (R6:0/10). The source bit come from I:000/10 and shifts into the first bit position, bit 16 of bit file B3.

After the processor completes the shift operation in one program scan, when the rung control bit I:000/03 goes false, the instruction resets the status bits EN, ER (if set), and DN.

For wraparound operation, the source address is assigned the same address as the highest (outgoing) bit address.

The structure of a bit shift right instruction is shown in Figure 8-9. This figure illustrates a BSR instruction with its parameters of file, control, bit address, and length. The instruction works in the same manner as a bit shift left instruction except the bits move to the right.

In the example in Figure 8-9, when the rung containing the BSR instruction goes from false to true, the processor sets the EN bit. Then the processor shifts 16 bits (length = 16) in bit file B3, starting with bit 31, to the right (to a lower bit number) one bit position. The last bit shifts out at bit position 16 into the control register UL bit (R6:0/10).

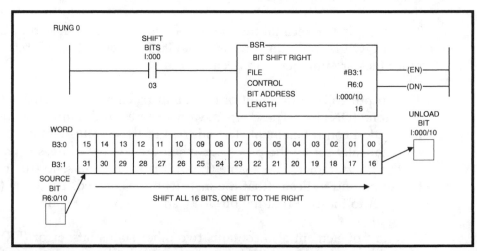

**Figure 8-9. An example of bit shift right (BSR) instruction.**

The source bit come from I:000/10 and shifts into the highest bit position, bit 31 of bit file B3.

After the processor completes the shift operation in one program scan, when the rung control bit I:000/03 goes false, the instruction resets the status bits EN, ER (if set), and DN.

For wraparound operation, the source address is assigned the same address as the highest (outgoing) bit address.

## First In-First Out (FIFO) Instruction

There are two FIFO instructions: the FIFO Load uses the mnemonic FFL and the FIFO unload uses the mnemonic FFU. These two instructions, FFL and FFU, are used in pairs to store and retrieve data in a prescribed order. When used in pairs, these instructions establish an asynchronous shift register or stack.

The two FIFO instructions, load and unload, must use the same file and control addresses, length, and position values as shown in the example of Figure 8-10.

This figure illustrates the FFL and FFU instructions with their parameters of source, destination, FIFO, control, length, and position.

The *Source* is the address that stores the "next in" value to the stack. The FIFO load instruction (FFL) retrieves the value from this address and loads it into the next word in the stack.

The *Destination* is the address that stores the value that exists from the stack.

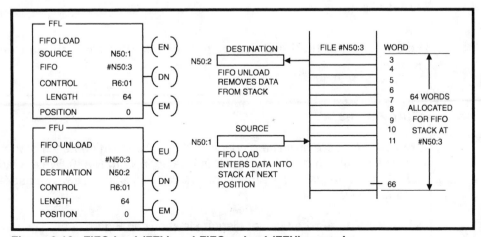

**Figure 8-10. FIFO load (FFL) and FIFO unload (FFU) example.**

The *FIFO* is an indexed address of the stack. The same FIFO address is used for the associated FFL and FFU instructions.

The *Control* is the address of the control structure (48 bits, or three, 16-bit words) in the control area of memory. The control structure stores the instruction's status bits, stack length, and next available position (pointer) in the stack.

The *Length* specifies the maximum number of words in the stack. The length is addressed by adding the mnemonic .LEN to the control address (R register).

The *position* indicates the next available location where the instruction loads data into the stack. The position is addressed by adding the mnemonic POS to the control address (R register). Normally the position value is 0, unless the programmer wants the instruction to start at an offset at power-up.

The output coils to the right of the FIFO instructions are the enable load (EN), enable unload (EU), empty (EM), and done (DN) bits. These bits have the same word address as the instruction control element (R). The address of these status bits is automatically set by the processor when the control address is entered by the programmer. The enable load (EN) bit used in FFL instruction is set by a false-to-true rung transition and indicates that instruction is enabled. The enable unload (EU) bit used in FFU instruction is set by a false-to-true rung transition and indicates the instruction is enabled. The done (DN) bit is set to indicate that the stack is full. The DN bit inhibits loading the stack until there is room. Empty (EM) is set by the processor to indicate that the stack is empty. The FIFO unload command should not be enabled if the EM bit is set.

The operation of the FIFO instructions can be illustrated by an example problem.

---

**EXAMPLE 8-3**

**Problem:** An assembly line produces 100 machine parts on each production run. The serial number for each part is located in word N40:0. Write a ladder logic program to place the serial numbers into File N50:3. Download serial numbers to word N40:1.

**Solution:** The required ladder diagram is shown in Figure 8-11.

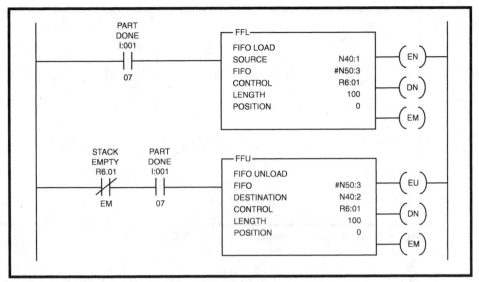

**Figure 8-11. FFL and FFU ladder diagram.**

## Sequencer Instructions

The sequencer instructions are typically used to control automatic assembly machines that have a consistent and repeatable operation. There are three common sequence instructions: sequencer input, sequencer output, and sequencer load.

The sequencer instructions are generally used to transfer data from the memory to discrete output modules for the control of sequential process operations or sequential batch operations (sequencer output). They are also used to compare I/O word data with data stored in tables so that process operating conditions can be examined for control and diagnostic purposes (sequencer input). The instruction is also used to transfer I/O word data into the memory (sequencer load).

The sequencer instructions can conserve program memory because they can monitor and control multiples of 16 discrete outputs at a time in a single rung of logic.

The sequencer input instruction (SQI) and the sequencer output instruction (SQO) are used in pairs to respectively monitor and control a sequential operation.

Figure 8-12 illustrates the three sequencer instructions. The parameters used in these instructions are file, mask, source, destination, control, length, and position.

The *file* is the indexed address of the sequencer file to or from which the instruction transfers data. Its purpose depends on the instruction.

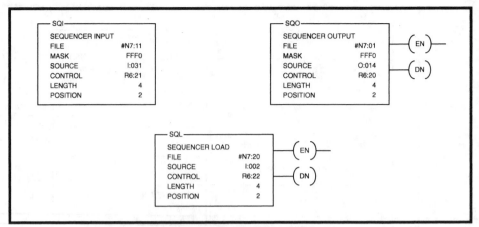

**Figure 8-12. Sequencer instructions SQI, SQO and SQL.**

The *mask* (for SQO and SQI) is a hexadecimal code or the address of the mask element or file through which the instruction moves data. Set (1) mask bits to pass data; reset (0) mask bits to prevent the instruction from operating on corresponding destination bits. Specify a hexadecimal value for a constant mask value. The programmer can store the mask in an element or file if he wants to change the mask according to a control application requirements.

The *source* (for SQI and SQL) is the address of the input element or file from which the instruction obtains data for its sequencer file.

The *destination* (for SQO, only) is the destination address of the output word or file to which the instruction moves data from its sequencer file.

Note that if the programmer uses a file for the source, mask, or destination of a sequencer instruction, the instruction automatically determines the file length and moves through the file step-by-step as it moves through the sequencer file.

The *control* is the address of the control structure in the control area (R) of memory (48 bits, or three, 16-bit words) that stores the instruction's status bits, the length of the sequencer file, and the instantaneous position in the file.

The programmer should use the control address with the mnemonic to address the following parameters:

Length (LEN) is the length of the sequencer file.
Position (POS) is the current position of the word in the sequencer file that the processor is using.

The *length* is the number of steps of the sequencer file starting at position 1. Position 0 is the start-up position. The instruction resets to position 1 at each completion.

Note that the address assigned for a sequencer file is step zero. Sequencer instructions use (length +1) words of data for each file referenced in the instruction. This also applies to the source, mask, and destination values if addressed as files.

The *position* is the word location in the sequencer file. The position value is incremented internally by SQO and SQL instructions.

To illustrate the operation of sequencer instructions, we will perform a sequencer output (SQO) example in Figure 8-13. In this example, the SQO instruction moves the data of the current step (2) through a mask to an output word that is connected to an A-B PLC5 output module in rack 1, I/O group 4.

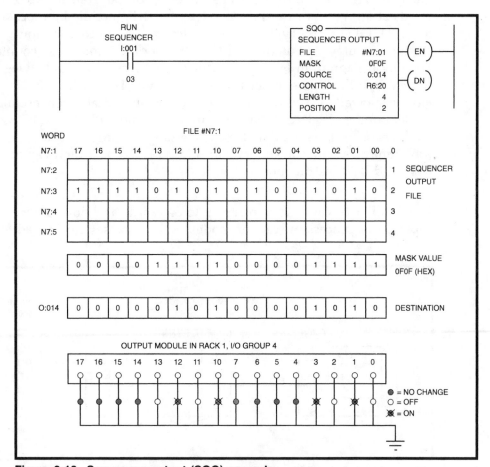

**Figure 8-13. Sequencer output (SQO) example.**

The SQO instruction steps through the sequencer file of 16-bit output words whose bits have been set to control the various output devices connected to the discrete output module shown. When the rung goes from false to true, the instruction increments to the next step (word) in the sequencer file #N7:1. The data in the sequencer file is transferred through a fixed mask (0F0F, HEX) to the destination address O:014. Current data is written to the destination element for every scan that the rung remains true.

## Block Transfer Instructions

In larger PLC systems, there is a requirement to transfer large blocks of data between supervisor PLC processors and remote processors on a high speed data communications network. In the A-B PLC5s, block transfers are performed using block-transfer write (BTW) and block-transfer read (BTR) instructions.

The basic A-B PLC5 processor in scanner mode can transfer up to 64 words at a time to or from a block transfer (BT) module in a local or remote I/O) chassis. Typical BT modules are thermocouple input modules, analog I/O modules, BCD I/O modules, and pulse counters. The block diagram in Figure 8-14 illustrates block transfers between a supervisory PLC5 in scanner mode to a remote I/O rack with BT modules installed. The remote I/O rack has an A-B adapter module, model number 1771-ASB that communicates internally to the BT modules using the rack data bus. The addapter (processor) also communicates with the supervisory PLC5 on the data highway using BTW and BTR instructions.

The A-B PLC5s can also transfer up to 64 words at a time between a supervisory process in scanner mode and a processor configured for

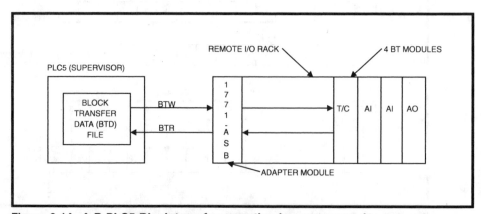

**Figure 8-14. A-B PLC5 Block transfer operation in scanner mode.**

adapter mode as shown in Figure 8-15. In this application, both processors simultaneously execute the opposite block transfer instruction.

Figure 8-16 shows a bidirectional alternating block transfer example. Using rungs like this example ensures that the block transfer requests are executed in the order in which they were sent to the queue. The processor alternates between the BTR and BTE instructions in the order in which they are scanned by virtue of the enable bit conditions in the rungs. Using the normally closed enabled bits from the two block transfer functions prevents the read and write block transfer instructions from queuing simultaneously. In this example, preconditioned normally open instructions are used to the left of the two enable bit contacts. These precondition bits allow time-driven or event-driven transfers.

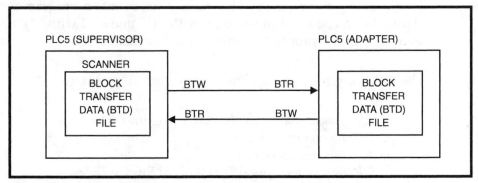

**Figure 8-15. A-B PLC5 block transfer operation in adapter mode.**

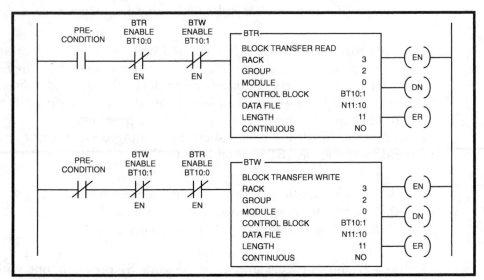

**Figure 8-16. Example bidirectional alternating block transfer.**

## Structured Text Language

Structured text (ST) is a high level language designed for complex automation processes. This language is used mainly to implement complex procedures that cannot be easily expressed with graphical languages. It is one of the IEC 1131-3 standard PLC languages. ST is also the default language for the description of the actions within the steps and conditions attached to the transitions of another IEC 1131-3 standard language, sequential function chart.

An ST program is a list of programming statements. Each statement ends with a semi-colon (";") separator. Names used in the source code, such as variable identifiers, constants, language keywords, are separated with inactive separators (space character, end of line, or stops) or by active separators, which have a well-defined significance (for example, the ">" separator indicates a "greater than" comparison. Programming comments can be freely inserted into a program list for clarity, but a comment must begin with "(*" and end with "*)". Each statement in the program terminates with a semi-colon (";") separator.

The following are basic statement types for ST programs:

1. *Assignment* statement (variable: = expression;),
2. *Subprogram or function* call,
3. *"C" function block* call,
4. *Selection* statements (IF, THEN, ELSE, CASE, etc.),
5. *Iteration* statements (FOR, WHILE, REPEAT, etc.),
6. *Control* statements (RETURN, EXIT, etc.), and
7. *Special* statements for links with other languages such as SFC.

Inactive separators may be freely entered between active separators, constant expressions, and identifiers to improve readability. The ST inactive separators are *space* (blank character), *tabs*, and *end of line* characters. Unlike line-formatted languages such as instruction lists, end of lines may be entered anywhere in the program. The programming tips listed below should be followed when using inactive separators to increase ST program readability:

1. Do not write more than one statement on one line,
2. Use tabs to indent complex statements, and
3. Insert comments to increase readability of lines.

Figure 8-17 shows two examples of the same ST program, one with low readability and the other with high readability.

```
LOW READABILITY ST PROGRAM:              HIGH READABILITY ST PROGRAM:

imax:=max_ite;cond:=X12;                 (*imax: number of iterations*)
if not(cond(*alarm*))                    (*i: FOR statement of index*)
then return;end_if;                      (*cond: process validity*)
for i (*index*) :=1 to max_ite
do if i<>2 then SPcall( );               imax:=max_ite;
end_if;end_for;                          cond:=X12;
(*no effect if alarm*)                   if not(cond(*alarm*))
                                         then return;
                                         end_if;

                                         (*process loop*)
                                         for i:=1 to max_ite do
                                         if i<>2 then
                                         SPcall( );
                                         end_if:
                                         end_for;
```

**Figure 8-17. ST program readability examples.**

## ST Expressions and Parentheses

ST expressions combine operators and variables or constant operands. For each single expression (combining operands with one ST operator), the type of the operands must be the same. The following are examples of valid and invalid expressions:

1. Expression: (boo_var1 AND boo_var2), this expression is valid because we have two Boolean variables (boo_var1 and boo_var2) being operated on by a Boolean AND operand, so that all items in the expression are the same type.
2. Expression: NOT(boo_var1), this is a valid expression because we have a Boolean variable (boo_var1) being operated on by a Boolean NOT operand.
3. Expression: (1s23 + 1.78), this is not a valid expression because there is a mixture of types used.

Parentheses are used to isolate parts of the expression, and to explicitly order the priority of the operations. If no parentheses are used in a complex expression, the operational sequence is implicitly given by the default priority between ST operators. For example, $2+3*6$ equals $2+18=20$ because the multiplication operator has a higher priority than the add operator. However, if parentheses are used on the same equation we have $(2+3)*6$ equals $5*6=30$ because priority was given by the parenthesis.

# Function Block Diagram Language

The functional block diagram (FBD) is a graphical language. It allows the programmer to build complex control procedures by taking existing functions from the FBD library and wiring them in a graphic diagram area. An FBD diagram describes a relationship or function between input and output variables. A function is described as a set of elementary function blocks as shown in Figure 8-18. Input and output variables are connected to blocks by connection lines. Typical input variables are temperature, pressure, level, position, and flow measuring devices or switches. Some typical output variables are motor starters, electric relays, contactors, and solenoid valves. An output of a function block might also be connected to an input of another block to perform logic or control functions. An input might also be a constant expression used in a control algorithm.

An entire function operated by a FBD program is built with standard elementary function blocks from the FBD library. Each elementary function block has a fixed number of input connection points and a fixed number of output connection points. For example, the Boolean AND function block shown in Figure 8-19 has two inputs and only one output. A function block is represented by a single rectangle with elementary function blocks such as the three Boolean AND functions shown inside the function block of Figure 8-18.

The inputs are connected on its left border. The outputs are connected on its right border. An elementary function block performs a single function between its inputs an its outputs. For example, the elementary function block shown in Figure 8-19 performs the Boolean AND operation on its two inputs and produces a result at the output. The

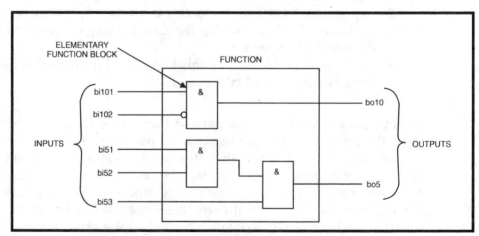

**Figure 8-18. FBD typical "function" format.**

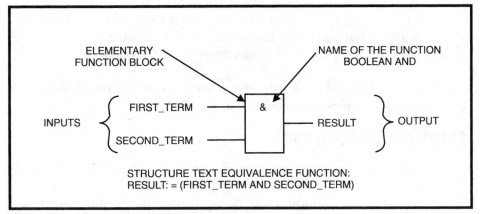

**Figure 8-19. Typical elementary function block.**

name of the function to be performed by the block is written in its symbol.

Each input or output of a block has a well-defined type. Input variables of a FBD program must be connected to input connection points of function blocks. The type of each variable must be the same as the type expected for the associated input. For example, an input to a Boolean function is expecting a logical one or zero and not an analog input. An input for an FBD diagram can be a constant expression, any internal or input variable, or an output variable.

Output variable of a FBD program must be connected to output connection points of function blocks. The type of each variable must be the same as the type expected for the associated block output. An output for an FBD can be any internal or output variable, or the name of the program (for subprograms only). When an output is the name of the currently edited subprogram, it represents the assignment of the return value for the subprogram (returned to the calling program).

Inputs and outputs of the function blocks are wired together with connection lines. Single lines may be used to connect two logical points of the diagram as follows:

1. An input variable and an input of a function block.
2. An output of a function block and an input of another block.
3. An output of a function block and an output variable.

The connections are oriented, meaning the line carries associated data from the left extremity to the right extremity. The left and right extremities of the connection line must be the same type.

Multiple right connections can be used to broadcast information or data from its left extremity to each of its right extremities. All of the extremities of the connection must be of the same type.

The FBD library consists of a large number of predefined standard operators, function blocks, and functions.

## Standard FBD Operators

Listed are standard function blocks of the IEC 1131-3 FBD language.

1. Data manipulation: Assignment, and Analog negation.
2. Boolean operations: Boolean AND, Boolean OR, and Exclusive OR.
3. Arithmetic operators: Addition, Subtraction, Multiplication, and Division.
4. Logic operations: Analog bit to bit AND mask, OR mask, and Exclusive OR mask.
5. Comparison tests: Less than, Less than or equal to, Greater than, Greater than or equal to, Is equal to, and Is not equal to.
6. Data conversion: Convert to Boolean, convert to integer analog, convert to real analog, convert to timer, and convert to message.
7. Other message concatenation: System access, operate I/O channel.

## Standard FBD Function Blocks

The FBD languages have the largest and most diverse set of functions of any PLC language. Table 8-3 lists the standard FBD function blocks. There are five Boolean types: set dominant bistable, reset dominant bistable, rising edge detection, falling edge detection, and semaphone. The FBD language includes the normal counters found in most PLC languages: Up counter, Down Counter and Up-Down Counters. The FBD timers are also the standard types such as On-delay timer and Off-delay timer but it also includes a Pulse Timer. The FBD language also includes two Analog types: Real and Integer as shown in Table 8-2. There are two types of signal generation in the FBD language: blinking Boolean signal and signal generator.

# Sequential Function Chart Language

The SFC language is used to describe operations of a sequential process. It uses a simple graphic representation of the different steps of a process, and conditions that enable the change of active steps. The SFC language is the core of the IEC 1131-3 standard. The other languages usually describe the actions within the steps and the logical conditions for the transitions.

| Type | Name | Description |
|---|---|---|
| Booleans | SR | Set dominant bistable |
| | RS | Reset dominant bistable |
| | RS_Trig | Rising edge detection |
| | F_Trig | Falling edge detection |
| | SEMA | Semaphore |
| Counters | CTU | Up counter |
| | CTD | Down counter |
| | CTUD | Up-down counter |
| Timers | TON | On-delay timer |
| | TOF | Off-delay timer |
| | TP | Pulse Timer |
| Integer analogs | CMP | Full comparison |
| | StackInt | Stack of integer analogs |
| Real analogs | AVERAGE | Running average over N samples |
| | HYSTER | Boolean hysteresis on diff. reals |
| | LIM_ALRM | HI/LOW limit alarm with hysteresis |
| | INTEGRAL | Integration over time |
| | DERIVATIVE | Differentiation over time |
| | PID | PID control |
| Signal generation | BLINK | Blinking Boolean signal |
| | SIG_GEN | Signal generator |

**Table 8-3. Standard FDB Function Blocks.**

An example of a sequential function chart program is shown in Figure 8-20 to explain the symbols used in a typical program. The top of the

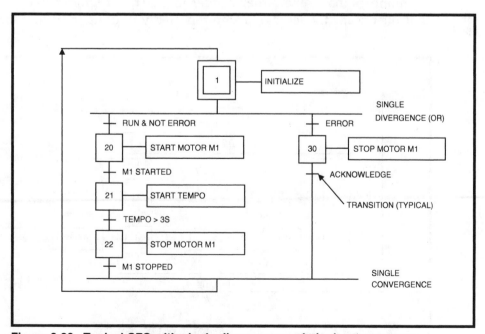

**Figure 8-20. Typical SFC with single divergence and single convergence.**

program contains a step block that is the *initial step*, where the programmable controller starts function chart execution and returns to this step from the end of program unless directed otherwise by the program logic. This block is identified by a double-sided box.

The *step* block is the function chart's basic unit and contains ladder logic for each independent stage of the process or machine operation. It is identified by a single-sided box.

The *transition* is the logic condition that the processor checks after completing the active step. When the transition logic is true, the step preceding the transition is disabled, and the step following it becomes active. The transition is normally a single logic rung, identified by a short horizontal line below its corresponding step (see Figure 8-20).

The *OR path* is identified by a single horizontal line at the beginning and end of a logic zone as shown in Figure 8-20. The processor selects one of several parallel paths, depending on which transition goes true first.

The *AND path* is identified by a horizontal double line at the beginning and end of a zone as shown in Figure 8-21.

In sequential function chart programs, steps and transitions are arranged in series and parallel paths, and they are numbered with the file numbers that contain their ladder logic. The programmable controller scans the logic of a step repeatedly until its transition logic

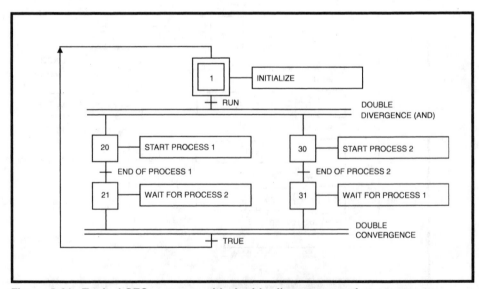

Figure 8-21. Typical SFC program with double divergence and convergence.

goes true. Then the program scan moves to the next step or steps, and the previously active step is turned off.

There are three basic rules for standard sequential function chart programs. The first is that the *initial* step is always activated at start-up. When restarting from the beginning of the chart and on subsequent passes through the flowchart program, the programmer does have the option of restarting from the beginning of the last active step(s), following or changing the programmable controller's mode from test or run to program.

The second rule is that a transition is tested after its associated step, and operations pass from one step to the next through a transition when the transition goes true.

The third rule is that after a true transition, the processor scans the step once more to reset all timer instructions and then executes the next step. This extra processor scan is called postscan. It is important to note that the processor never postscans a transition file, so timers probably should not be used in transition files.

The processor scans a sequential function chart program from left to right and top to bottom. When the sequential function chart program scan encounters active parallel steps, it executes the ladder logic in the leftmost step first, then moves to the ladder logic in the next parallel step across the screen from left to right.

## SFC Application

To illustrate an SFC application, consider a simplified example of a semiautomatic punch, shown in Figure 8-22. We assume the punch starts in the raised position or top position. When the operator depresses the start pushbutton, the punch is lowered and it pierces the metal part at the lowest or bottom position. The cycle is completed

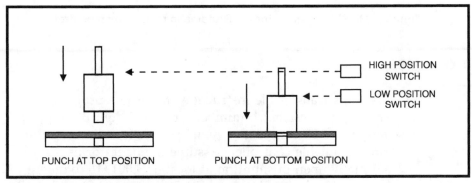

**Figure 8-22. Semiautomatic metal punch example.**

when the control system raises the punch back to the top position and the operator removes the punched metal part and inserts a new piece of metal for the next operation.

In the SFC language representation of the simple semiautomatic punch control system shown in Figure 8-23, the system has three states or steps. The first state is the punch at rest in the raised position, the second step is the punch descending, the final step is the punch ascending.

When the punch and associated control system are activated, the programmable controller places the system in the Wait state (step 1) and waits for the first transition logic to be satisfied. If the start PB is depressed AND the punch is in the top position, the first transition is true and step 2 is activated, so the punch moves down. At the same time, the programmable controller turns OFF the first state.

When the punch activates the bottom limit switch, the second transition is true, and the punch raises. Finally, when the punch reaches the top, the final transition is set and the program returns to the Wait state.

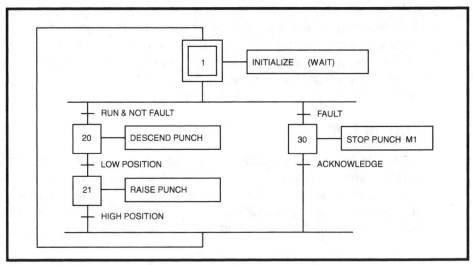

**Figure 8-23. SFC program for semiautomatic metal punch control.**

## EXERCISES

8.1   Use data transfer file instructions to write a program to transfer process batch numbers from a 3-digit BCD input module to 20 consecutive memory locations in a programmable controller. Assume the operator uses a normally open pushbutton at location I:001/16 to enter each BCD number. Use an LED indicator at output location

O:002 to display the batch number and an another LED at O:003 to display the data selected at the batch step.

8.2   Write a ladder logic program to transfer 20 words of data starting at memory word N60:0 to a block of memory starting at word N70:0, if input bit I:000/13 is true.

8.3   Write a software program to transfer 10 words of process data from consecutive memory locations starting at word N70:10 to a panel-mounted LED display. Assume the operator uses a control panel pushbutton at address I:000/10 to request process data. Use a second LED display to indicate the memory location of the data to the operator. The LED displays are connected to the digital output modules at memory locations N70 and N71.

8.4   Design a software program to read data every 0.02 second from an 8-channel analog input module in rack 1, module group 5. Send the same data to an 8-channel analog output module in rack 0, module group 7 right after the data read instruction has been completed.

8.5   Use file instructions to write a program to transfer process set point numbers from a 3-digit BCD input module to 8 consecutive memory locations in an A-B PLC5. Assume the operator uses a normally open pushbutton at location I:000/00 to enter each BCD number.

8.6   Write a program to store the last ten temperature readings from the seventh input channel of an 8-channel analog input module located in rack 0 and module group 2.

## BIBLIOGRAPHY

1.   Jones, C. T., and Bryan, L. A., *Programmable Controllers: Concepts and Applications,* International Programmable Controls, Inc., First Edition, 1983.

2.   Modicon 584, *Programmable Controllers: User's Manual,* Gould Modicon Division, January 1982.

3.   PLC2/30, *Programmable Controller: Programming and Operations Manual,* Allen-Bradley, Publication 1772-6.8.3, 1984.

4.   Hughes, T. A., *Basics of Measurement and Control,* Instrument Society of America, Second Edition, 1995.

5.   Gilbert, R. A., and Llewellyn, J. A., *Programmable Controllers: Practices and Concepts,* Industrial Training Corporation, 1985.

6.   Lloyd, M., *Grafcet Graphical Functional Chart Programming for Programmable Controllers,* Measurement & Control Magazine, September, 1987 edition.

7.   Christensen, J. H., *Programmable Controller Users and Makers to Go Global with IEC1131-3, I&CS Magazine,* October 1993.

# 9

# Data Communication Systems

## Introduction

The purpose of communications is to transfer information from one point to another or from one system to another. In process control, this information is called *process data* or, simply, *data*.

An understanding of data communications is essential for the proper application of programmable controllers to process control and data collection. This chapter will provide a basic understanding of data communications terminology and concepts and their application to programmable controller systems.

## Basic Communications

Data is transmitted through two types of signals: baseband and broadband. In a *baseband system*, a data transmission consists of a range of signals sent on the transmission medium without being translated in frequency. A telephone call is an example of a baseband transmission. A human voice signal in the 300- to 3000-Hz range is transmitted over the phone line in the same frequency range. In a *baseband system*, there is only one set of signals on the medium at a time.

A broadband transmission consists of multiple sets of signals. Each set of signals is converted to a frequency range that will not interfere with other signals on the medium. Cable television is an example of broadband transmission.

As shown in Figure 9-1, three basic components are required in any data communication system: the *transmitter* to generate the information, the *medium* to carry the data, and the *receiver* to detect the data.

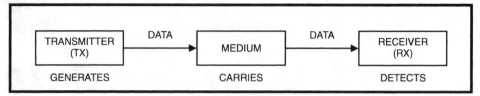

**Figure 9-1. Basic communication system components.**

The medium can be divided into more than one channel. A channel is defined as a path through the medium that can carry information in only one direction at a time.

Figure 9-2 shows an example of a multichannel medium with n channels. Probably the most common example of multi-channel communication system are cable television systems.

Physical transmission media fall into four general categories: multiconductor cable, twisted pair cable, coaxial cable, and fiber optic cable. Multiconductors consist of two or more insulated electrical conductors inside a plastic-covered cable. A typical example is the parallel printer cable used between a personal computer and a printer. Twisted pair cables consist of two electrical conductors, each covered with insulation. The two wires are twisted together to ensure that they are both equally exposed to electrical interference signals in the environment. Since the wires are carrying current in opposite directions, the electrical interference will tend to cancel out in the cable. Twisted pair is the most common cable used in programmable controller systems. It is the least expensive transmission medium and provides adequate electromagnetic interference (emi) immunity.

A coaxial cable consists of an electrical conductor surrounded first by insulating material and then by a tube-shaped metal braid conductor. In most cases, the entire cable is covered by an insulator. The round conductor in the center of the cable and the circular outer tube conductor are coaxial in that they share the same central axis.

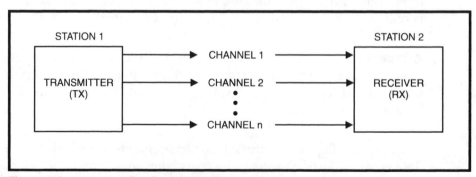

**Figure 9-2. n-communication channels.**

Coaxial cables are used in both baseband and broadband communications systems. A variety of cable types are used in process control applications, ranging from flexible types similar to coaxial cables used on home TV sets to heavy, rigid cables that require special installation procedures.

Coaxial cables are generally used in process automation applications where long communications runs of over 1000 feet are involved and improved emi immunity is required. They find very extensive use in plant-wide communication networks.

The fiber optic cable consists of fine glass or plastic fibers. At one end, electrical pulses are converted into light by a photo diode and sent down the fiber optic cable. At the other end of the cable, a light detector converts the light pulses back to electrical signals. The light signals can travel only in one direction, so two-way transmission requires two separate fiber cables. A fiber optic cable is normally the same size as a twisted pair cable, and it is immune to electrical noise.

The cost of the optical fiber cables is about the same as coaxial cables. However, the use of fiber optics in programmable controller applications has been limited by the high cost of connectors and lack of industrial standards.

Communication can also be described by the number of channels used to effect the information flow. The three common methods of data transmission are unidirectional, half-duplex, and duplex.

# Unidirectional Communications

In unidirectional communications, a single channel is used and communication is only one way (i.e., from transmitter to receiver), so the receiver can never respond.

In unidirectional communication, it is not possible to send errors or control signals from the receiver station, because the transmitter (TX) and the receiver (RX) are each dedicated to performing one function. A typical example of unidirectional communications in process control would be a weigh scale in the field sending data to a programmable controller in a central control room.

## Half-Duplex Communications

Two-way communication allows the receiver to verify that the data was received. One kind of two-way communication is called half-duplex. In half-duplex, a single channel is used and communication is two-way, but communication can occur only in one direction at a time (see Figure 9-3). In this configuration, the receiver and transmitter alternate functions, so communication occurs in one direction at a time on the single channel. A circuit in the station communication module acts as an OR gate to place the unit in transmit or receive mode.

## Full-Duplex Communication

Two-way communications where data can flow in both directions at the same time is called *full-duplex communications*. In this case, two physical paths are required, so information can flow from both directions simultaneously, as shown in Figure 9-4.

# Transmission Methods

*Transmission* involves moving data between a receiver and a transmitter. The two methods of transmission are *parallel* and *serial*.

## Parallel Transmission

In parallel transmission, all bits are transmitted at the same time, and each bit of information requires a unique channel. For example, if we transmit an eight-bit ASCII character, eight channels are required.

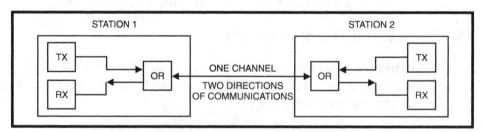

**Figure 9-3. Half-duplex communication.**

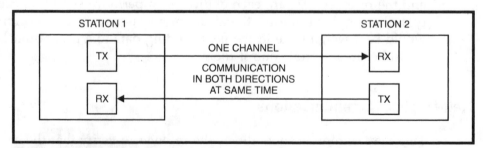

**Figure 9-4. Full-Duplex communication.**

Figure 9-5 demonstrates the transmission of the ASCII character using odd parity. The term "parallel" refers to the position of the bits of the character and the fact that the characters are transferred as a group of bits in parallel. The use of a group of channels results in a high data transfer rate.

There is a problem with bit synchronization delay in parallel data transfer. Synchronization is the process of making two or more activities happen at the same rate and time. If we send a string of characters on a long parallel cable, the difference in impedances in the cable wires can cause bit synchronization decay, as shown in Figure 9-6.

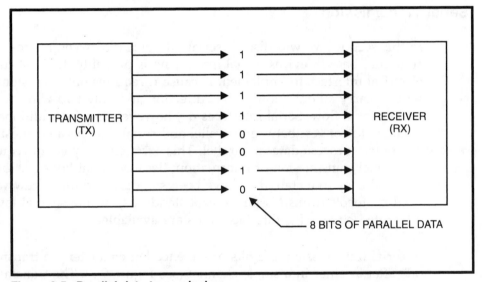

**Figure 9-5. Parallel data transmission.**

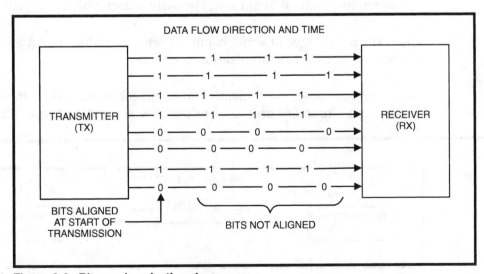

**Figure 9-6. Bit synchronization decay.**

To overcome this problem, the distance between the receiver and the transmitter must be kept short to avoid transfer errors caused by bit synchronization decay. For this reason, parallel transmission is used only in closely connected data systems where high speed is required. For example, data transfer on data and address buses in a microprocessor is parallel transmission. However, in process control systems the data transfer between systems such as personal computers and programmable controllers is normally serial. For this reason, we will limit our discussion in the remaining parts of this chapter to serial data transmission.

## Serial Transmission

In this section, we will discuss serial interfacing between microcomputer-based devices. Serial interfacing is probably the least complex electrical interface to implement because it requires only one signal wire to carry all data flow in one direction and only two wires for bidirectional flow. Serial interfaces do, however, require additional logic circuitry to convert between parallel and serial data because most computers process data in parallel. This extra circuitry is not required in parallel transmission of information. Because serial links are low in cost and easy to install, they have been standardized into a few widely used protocols (transmission control standards), where several low cost integrated circuit (IC) interface chips are available.

In serial transmission, the bits of the encoded character are transmitted one after another in a single channel (see Figure 9-7). The transmission takes the form of a bit stream that the receiver must assemble into characters (normally 8 bits) using specially designed ICs.

The main advantages of serial communications are lower cost and the elimination of byte synchronization problems.

Controlling the speed of transmission is critical in serial data transfer. A common measurement unit used to describe serial data speed is *bits per sec (bps)*.

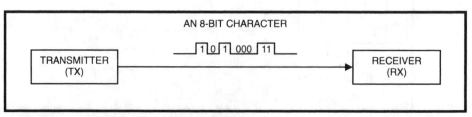

**Figure 9-7. Serial 8-bit character transmission.**

The process of bit synchronization is required for serial data transfer. A timing circuit at the transmitter places transmitted bits on the channel at fixed intervals determined by the selected communication rates. Some typical communication rates used are 2.4K, 4.8K, 9.6K, 14.4K, and 28.8K bps.

The transmitter must send bits at the same rate the receiver is set to accept bits. For example, if a personal computer is transmitting serial data at 9600 bps to a programmable controller system, the communications module on the programmable controller must be set at a communications rate of 9600 bps.

Character synchronization is used to determine which eight consecutive bits in a data stream represent a character. The receiver must recognize the first data bit and count bits until the character or byte is complete. Various forms of character synchronization are used in data communications. Sometimes each byte is framed with special bits to indicate start and stop.

## Signal Multiplexing

A single line can be used to carry more than one signal by using signal multiplexing. To understand multiplexing, we must first define the term "bandwidth." *Bandwidth* describes the signal-carrying capability of a communications channel or line. Bandwidth is defined as the difference in cycles per second between the lowest and highest frequencies a channel can handle without a certain amount of signal loss.

Multiplexing allows one line to serve more than one receiver by creating slots or divisions in the line. The equipment used to obtain multiplexing is called a multiplexer or MUX (see Figure 9-8).

The commonly used signal multiplexing methods are frequency division, time division, and statistical.

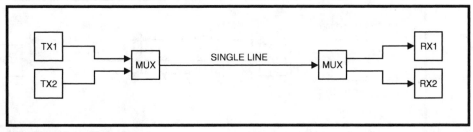

**9-8. Signal multiplexing.**

## Frequency Division Multiplexing

In frequency division multiplexing, each channel is assigned its own individual frequency range. Frequency division is normally used to combine a group of low-speed sources onto a single voice line. A typical application of frequency division communications is shown in Figure 9-9.

## Time Division Multiplexing

Time division multiplexing uses time periods to assign channel space and allot the available bandwidth. In Figure 9-10, two time periods (Period 1 and Period 2) are used to assign each channel its own individual time slot in the data stream. In this type of multiplexing, no other channel can use the time slot, so there is some wasted bandwidth when a station is not transmitting data in its time slot. However, one transmission line can support many data streams. It is normally used where there is a need to combine a number of relatively low speed data transmissions onto a single high speed line.

## Statistical Multiplexing

Statistical multiplexing is an enhancement of time division multiplexing designed to reduce the wasted bandwidth when data is being transmitted by a station. This enhancement process is sometimes called data concentration. Statistical multiplexing is generally used where a large number of data entry terminals require only a brief or occasional data

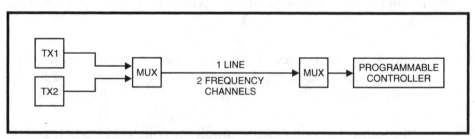

Figure 9-9.  Frequency division multiplexing.

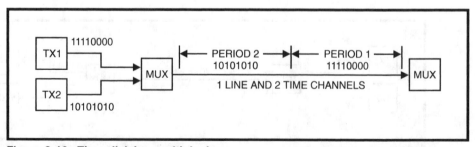

Figure 9-10.  Time division multiplexing.

transfer. If the use of the terminals becomes intensive, the operators will experience long time delays, so this method must be applied with caution.

# Error Control and Checking

Error control and checking is the process used to detect any discrepancy between transmitted data and received data in a communication system. Errors detected at the receiver are either corrected or retransmitted. The common error control methods used in programmable controller systems are (1) echo check (Echoplex), (2) vertical redundancy check (VRC), (3) longitudinal redundancy check (LRC), and (4) cyclical redundancy check (CRC).

## Echo Check

Echo check is used in two-way communications systems (Figure 9-11) to check the accuracy of transmitted data.

The receiver (RX) in station 2 sends every character received back to the originating terminal (station 1). Station 1 compares the echoed data with the information it sent. The comparison is performed by an electronic circuitry in the transmitter/receiver and if a transmission error is detected, the information is retransmitted a fixed number of times to obtain an error free transmission.

## Vertical Redundancy Check (VRC)

In the vertical redundancy check process, parity bit checking is used to detect the change of a single bit. In this method, a single bit is added to a character string to create either an odd or an even number of 1 bits. Parity bits are often called "redundant" because they can be removed without loss of data. If even parity is used, a 1 bit is added to a character string to make the total of 1 bits even. For example, the ASCII letter "A" is given by 0000001; if even parity is used, a 1 bit is added to the binary string before transmission (i.e., A = 10000001).

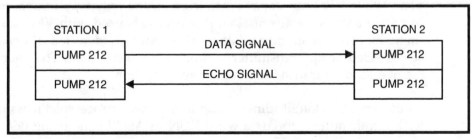

Figure 9-11. Echo check.

In even parity VRC transmission, the receiver would detect an error when it receives a character that contains an odd number of 1 bits. In this method, after a parity error is detected, the receiver would request retransmission of the data.

An example problem will help to explain the concept of parity.

---

**EXAMPLE 9-1**

**Problem:** Generate the ASCII code with even parity for the word STOP.

**Solution:** Using the ASCII table in Chapter 2, we first obtain the binary code for each character as follows:

| Bits | 7 | 6 | 5 | 4 | 3 | 2 | 1 | Character |
|------|---|---|---|---|---|---|---|-----------|
| | 1 | 0 | 1 | 0 | 0 | 1 | 0 | S |
| | 1 | 0 | 1 | 0 | 1 | 0 | 0 | T |
| | 1 | 0 | 0 | 1 | 1 | 1 | 1 | O |
| | 1 | 0 | 1 | 0 | 0 | 0 | 0 | P |

Next, we count the number of 1 bits for each character. If the number is even, we set the parity (P) bit to 0 to maintain an even number of 1 bits in the binary string. On the other hand, if the number of 1 bits is odd, we set the parity bit to 1 to obtain an even number of 1 bits in each string as shown:

| Bits | P | 7 | 6 | 5 | 4 | 3 | 2 | 1 | Character |
|------|---|---|---|---|---|---|---|---|-----------|
| | 1 | 1 | 0 | 1 | 0 | 0 | 1 | 0 | S |
| | 1 | 1 | 0 | 1 | 0 | 1 | 0 | 0 | T |
| | 1 | 1 | 0 | 0 | 1 | 1 | 1 | 1 | O |
| | 0 | 1 | 0 | 1 | 0 | 0 | 0 | 0 | P |

---

## Longitudinal Redundancy Check (LRC)

Longitudinal redundancy check (LRC), or block check character (BCC), is a procedure that checks an entire horizontal line within a block of data for odd or even parity. This process is used in combination with VRC. While VRC can detect a single error, the only way to obtain corrected data is retransmission. When LRC is used with VRC, any single bit error in an entire data block is not only detected but can also be corrected at the transmitter without a retransmission. This increases the overall data transmission speed for a system.

To illustrate the longitudinal redundancy check process, let's assume we are transmitting the data word RUN in ASCII code using odd parity as shown in Figure 9-12.

Note that in Figure 9-12, an LRC odd parity character is sent by the transmitter. This character is generated by adding a parity bit at the end of the entire transmitted block so that an odd number of 1 bits is created for each longitudinal row of bits.

At the receiver, the LRC is calculated for the data bytes in the block, and it is compared with the transmitted LRC character. If they are not equal, the vertical parity bit reveals which byte is in error, and the LRC reveals which of the 8 bits is in error. Logic circuitry or software in the receiver is used to change the error bit to its opposite state, correcting the error. Only if there is more than one error in the block transfer must retransmission be requested by the receiver.

**EXAMPLE 9-2**

**Problem:** Calculate the LRC/VRC odd parity characters for the word PUMP.

**Solution:**

| Word | P | U | M | P | LRC Parity |
|------|---|---|---|---|------------|
| Bit 1 | 0 | 1 | 1 | 0 | 1 |
| Bit 2 | 0 | 0 | 0 | 0 | 1 |
| Bit 3 | 0 | 1 | 1 | 0 | 1 |
| Bit 4 | 0 | 0 | 1 | 0 | 0 |
| Bit 5 | 1 | 1 | 0 | 1 | 0 |
| Bit 6 | 0 | 0 | 0 | 0 | 1 |
| Bit 7 | 1 | 1 | 1 | 1 | 1 |
| Vertical Parity Bit | 1 | 1 | 1 | 1 | 0 |

## Cyclical Redundancy Check (CRC)

Cyclical redundancy check (CRC) is a method for checking error on an entire data block. In this method, a check character is transmitted at the end of each block. The transmitter calculates the check character from the data transmitted.

The receiver compares the transmitted CRC to its own CRC, calculated using the received block of data. If they are not equal, the receiver requests retransmission of the previous message block. If they are equal, the transmitted data is assumed to be error-free.

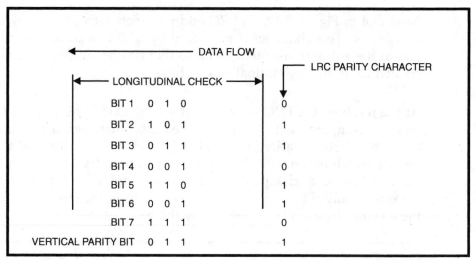

DATA FLOW

LRC PARITY CHARACTER

LONGITUDINAL CHECK

| | | | | | |
|---|---|---|---|---|---|
| BIT 1 | 0 | 1 | 0 | | 0 |
| BIT 2 | 1 | 0 | 1 | | 1 |
| BIT 3 | 0 | 1 | 1 | | 1 |
| BIT 4 | 0 | 0 | 1 | | 0 |
| BIT 5 | 1 | 1 | 0 | | 1 |
| BIT 6 | 0 | 0 | 1 | | 1 |
| BIT 7 | 1 | 1 | 1 | | 0 |
| VERTICAL PARITY BIT | 0 | 1 | 1 | | 1 |

**Figure 9-12. Longitudinal redundancy check example.**

Cyclical redundancy check is a very accurate method for detecting data transmission errors. It is the most common method used in programmable controller-based systems.

## Communication Protocols

Communications protocol is a set of rules specifying the format and control of transmission between two communication devices. The two main functions of protocol are *handshaking* and *line discipline*. Handshaking determines whether a circuit is available and makes sure the circuit is ready to transfer data. Line discipline performs the following functions: (1) receive and transmit information, (2) error control procedures, (3) sequencing of the message blocks, and (4) error checking.

To illustrate the concept of communications protocol, a typical line discipline sequence between a transmitter and receiver is shown in Figure 9-13.

The two basic transmission methods used to obtain line discipline are asynchronous and *synchronous*. In *asynchronous*, successive data appear in the data channel at arbitrary times, with no specific clock control governing the time delays between information. In *synchronous*, each successive datum in a data stream is controlled by a master data clock and appears at a specific interval of time.

Most synchronous and asynchronous systems deliver serial data in 8-bit characters. The asynchronous systems treat each character as an individual message, and the characters appear in the data stream at

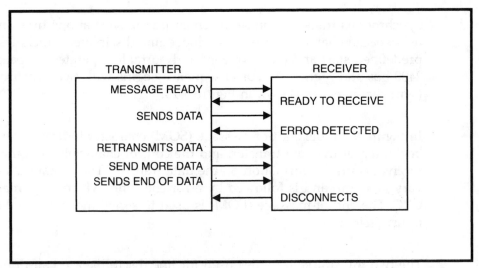

**Figure 9-13. Typical line discipline sequence.**

arbitrary relative times. However, within each character, the 8 bits are transmitted at a fixed predetermined clock rate such as 2400, 9600, 14.4K, 28.8K, and 33.6K bps. Hence, an asynchronous communications system is actually synchronous within a character and asynchronous between characters.

A typical bit stream for an asynchronous transmission is shown in Figure 9-14. Line discipline is achieved by framing every character transmitted with start/stop bits. The start bit is a O or "space" and the stop bits are a 1 or "mark."

Asynchronous transmission operates at random speeds and is almost always used with half-duplex protocols.

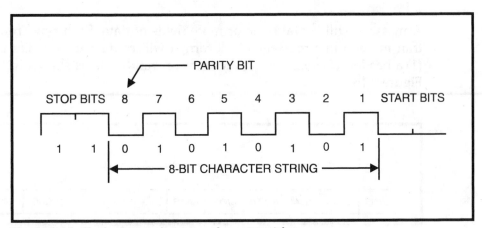

**Figure 9-14. Typical asynchronous character string.**

Synchronous transmission sends an entire message at one time and uses special character strings to achieve line discipline. It includes a predefined start and end sequence and normally operates at speeds of 2400 bps and higher. A typical sequence begins with two synchronous (syn) characters as shown in Figure 9-15.

In Figure 9-15, the start-of-message (SOM) character indicates the beginning of the data transfer, and the control character gives the receiver control information on the data transfer. The actual data can vary in length and is followed by error checking, which is normally CRC. The end-of-message (EOM) is used to signal the end of a data transmission.

The two transmission modes used for line discipline are half-duplex and full duplex protocols. Half-duplex protocol allows for communications in only one direction. In a 2-wire half duplex circuit, the stations must be able to switch automatically from transmit to receive. In a 4-wire full duplex circuit, each station has a dedicated receiver and transmitter circuit, but data cannot be transmitted and received at the same time. Full duplex protocol permits simultaneous transmissions in both directions. It falls back to half duplex operation if the communications link cannot support full duplex.

## Serial Synchronous Transmission

Synchronous communications depend on the data and protocol control information being assembled into a structured and predefined package or format. The format tells the computer or other equipment what to do with the message and how to do it.

A message will contain one or more fields of data. Each synchronous transmission in a message block format will contain three major parts: (1) a header, (2) text, and (3) the trace or trail block, as shown in Figure 9-16.

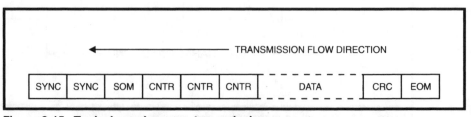

**Figure 9-15. Typical synchronous transmission.**

The *header* starts the message package and provides the capability for synchronization between the transmitter and receiver. It normally contains characters that identify the transmitting or receiving stations and provides communications routing data. The *text* contains the data characters to be transferred and consists of a single- or multiblock message. The *trace* or *tail* characters signal the end of a transmission block.

## Serial Synchronous Protocols

Different control and data formats are used by different protocols. A communications protocol can be defined as a fixed set of rules governing the format and control of inputs and outputs between two data transfer devices. In a standard data transfer system, a protocol governs the following: line control, framing, error control, and sequence control.

*Line control* is used to list the station that will transmit and the station that will receive in half-duplex communications. It is also required in a multipoint circuit (a circuit connecting 3 or more points on a common line). The *framing* function normally determines which 8-bit groups are the characters and what groups of characters are the messages. *Error control* detects transmission errors using various redundancy checks and normally corrects faulty messages. *Sequence control* function numbers the messages to eliminate duplication and avoid data loss. It also identifies any retransmitted messages.

There are basically three protocols in wide use today in process control communication systems:

1. BISYNC (Binary Synchronous Communications), an older protocol used in IBM equipment.
2. DDCMP (Digital Data Communication Message Protocol), a protocol used primarily by Digital Equipment Corporation (DEC)™ equipment.
3. (High-level Data-Link Control), probably the most widely used protocol.

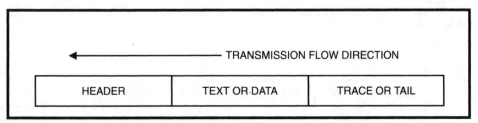

**Figure 9-16. Three elements of a typical synchronous transmission.**

Because BISYNC and HDLC protocols are the most widely used in programmable controller systems, we will limit our discussion to these two techniques. HDLC has evolved from two earlier standards, the ADCCP (advanced data communications control procedure) and the SDLC (synchronous data link control). Because of the close similarity of the three, we will cover only the HDLC protocol in this discussion. But first, we need to discuss the BISYNC protocol.

## BIYNC Protocol

BISYNC stands for BInary SYNChronous communications; it is a half-duplex character oriented protocol. The protocol has a very rigid format that uses special characters (ASCII or EBCDIC) to delineate the various fields of a message and to control the required protocol functions.

A typical BISYNC message, shown in Figure 9-17, consists of the following discrete parts: (1) two or more synchronizing characters (SYNC), (2) start of header (SOH), (3) header, (4) start of text (STX), (5) text, (6) an end of text (ETX), and (7) a trace or tail block (CRC).

The synchronizing (SYNC) characters are used to establish the correct timing between the transmitter and receiver. The number of SYNC characters varies with the different communication applications and networks. The start of header is a format control character that is transmitted just before the header character to identify the individual message control characters. The header is an optional character that normally contains routing or message priority information. The start of text (STX) is a special format control character that is transmitted before the first data characters; it indicates that the characters to follow are information. The text, of course, is the data that is being transmitted. The end of transmission block (ETB) is a format control character indicating the end of text and the beginning of the trace or tail (CRC). It is normally used to indicate the end of an intermediate text block.

The end of text (ETX) is also a special format control character that indicates the end of a text block and the start of the trace or tail

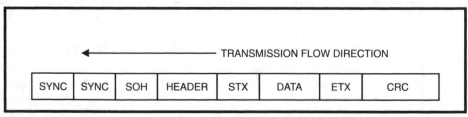

**Figure 9-17. A BISYNC transmission string.**

(CRC). The trace or tail block (CRC) detects and corrects errors in transmission. It depends on the information code being used, such as ASCII or EBCDIC, and has a block check character or a combination of checks.

If the trace or tail block is in ASCII, a VRC/LRC message check is performed. In EBCDIC transmission, normally no VRC/LRC check is performed, and the CRC is calculated on the entire message.

The BISYNC is a rather simple protocol, but several problems complicate it. The format of the protocol places special meaning on the ETX character. If the data block (or control information) contains this character among its data, the characters can be misinterpreted. For example, if the datum happens to be an 8-bit pattern identical to the ASCII pattern for ETX, this datum character could deceive the receiver into taking an end-of-block action when actually the message block has more characters to follow. The correct this problem, the protocol needs to distinguish specific patterns as data characters. The ability to treat control characters either as control information or as data is called *data transparency*.

BISYNC uses the control character DLE (data link escape) to obtain the required transparency. If a control symbol is to be treated as data, it is preceded by the DLE character. In other words, the receiver is warned by the receipt of a DLE to accept the next character as data and not to take any control action. The use of the DLE character is somewhat more complicated than described because of other special considerations. For example, to maintain transmission synchronization in the absence of data in the transmitter queue, the protocol provides for an automatic insertion of sync characters, which are ignored by the receiver. But it is possible that a sync character might be inserted between a DLE and a control character that follows it. This condition forces the receiver to interpret the sync character as if it were data, rather than accepting the next character as data. In this situation, the transmitter cannot simply send a sync character after a DLE. The transmitter must queue the DLE, and then send sync characters until both the DLE and its corresponding data character are ready. If message buffering is not available, the transmitter has to send a DLE and then stay idle on the line. In the situation, it should transmit in the idle state the characters DLE and SYNC, which the receiver can ignore as pairs of characters. Also note that since the DLE is a control character with special significance, the DLE character in the data transmission must be treated the same way the ETX is treated, and it must be preceded by a DLE character when transmitted over a communications link. An example problem will help to illustrate the BISYNC protocol.

**EXAMPLE 9-3**

**Problem:** Assemble the bit streams required to send the message M1 ON using 8-bit odd parity ASCII code and BISYNC protocol. Assume the header is given by 00000001 for transmitting station 1 and omit the check character (BCC or CRC) at the end of the transmission.

**Solution:** Use the ASCII code table to find the bits required as follows: SYNC = 00010110, SOH = 00000001, STX = 00000010, M = 11001101, 1 = 00110001, space = 00100000, 0 = 01001111, N = 11001110, ETX = 10000011.

## The HDLC Protocol

The main feature of HDLC protocol is that it opens and closes each message block or frame with start-frame and stop-frame characters (flags). A typical HDLC message is shown in Figure 9-18 and consists of six discrete parts as follows: (1) open flag, (2) address byte, (3) control byte, (4) data field, (5) a check field, and (6) a close flag. The open flag always consists of the same 8 bits (01111110); it is used to indicate the start of a transmission frame. This 8-bit sequence is never repeated again throughout the entire message until the close flag.

The address byte is the address of the transmitter on a command message or the receiver on a response message. The address byte consists of 8 bits allowing for 256 addresses, but 16 addresses is the maximum normally used since some bits are used for other functions.

The control byte has 8 bits and contains command or control response information. The data field, on the other hand, can contain any number of bytes and is the user's total data transmission. The data field normally uses EBCDIC, ASCII, BCD, or straight binary code.

The check field follows the data and precedes the closing flag. It contains a cyclical redundancy check (CRC) character that detects and in some cases corrects errors during transmission. To review, CRC functions as follows: the transmitter sends its computation to the

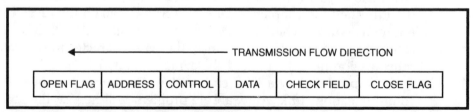

**Figure 9-18. An HDLC transmission string.**

receiver, which compares the transmitted computation with its own calculation. If equal, the data is assumed to be error-free, and, if unequal, the receiver may not accept the transmission and normally requests retransmission.

The close flag is the final transmission byte and has the same bit configuration as the open flag. It terminates the transmission frame and begins the next if it is available.

The synchronous nature of this protocol forces the transmitter to have data ready in a buffer at the beginning of a transmission block. If it is not ready and the system fails to produce the data in time for transmission, the transmitter will run out of information to transmit. The HDLC protocol does not have an idle character within a block, so the system must abort an entire transmission block when the transmitter runs out of data to send. The abort code is normally a sequence of eight 1s.

# Local Area Networks

A local area network (LAN) is a user-owned and operated data transmission system operating within a building or set of buildings. LANs allow a great number and variety of machines and processes to exchange large amounts of information at high speed over a limited distance. LANs connect communicating devices, such as computers, programmable controllers, process controllers, terminals, printers, and mass storage units within a single process or manufacturing building or plant.

LANs allow neighboring computers to share data resources, hardware resources, and software. For example, in a typical manufacturing facility, the process control system computers and a central computer used by purchasing might be tied together to speed up the ordering of raw material for the plant.

A typical LAN for an industrial facility is shown in Figure 9-19. It consists of three levels with three different types of networks. The highest level (Level I) is the information network used by groups like accounting and purchasing at the plant. This network has the highest speed communication (10 Mbps and higher) network because it must handle large amounts of data and information. The middle level (Level II) is used to perform the automation and control of the industrial plant. This network is generally slower (normally 100 Kbps to 5 Mbps) than the level I network because the data rates required for process and machine control are lower. The low level bus (Level III) is used to

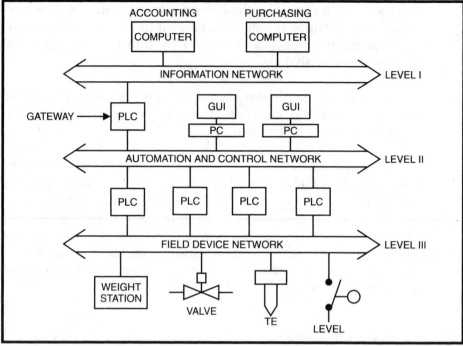

**Figure 9-19. Typical industrial plant LANs.**

connect programmable control and other controllers directly to the field devices such as weight, flow, pressure, level transmitters, and switches.

## LAN Topologies

There are three LAN topologies in common use: ring, bus, and star. The *ring* network topology shown in Figure 9-20 can operate using unidirectional transmission medium. However, most ring topology networks use bidirectional transmission, allowing messages to flow in

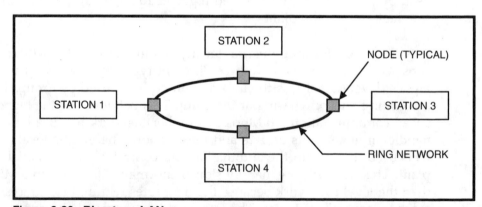

**Figure 9-20. Ring-type LAN.**

the most efficient manner. Each node decides whether to accept or pass on a message. This scheme is relatively easy to implement.

The *bus* network topology shown in Figure 9-21 requires a broadcast medium in which signals flow to all stations at all times. All the stations receive transmissions, even if they act only on some. The advantage of this scheme is that the stations connected to the bus perform no message routing because the bus is a broadcast type medium. This is the most common found in process control.

The *star* network shown in Figure 9-22 normally allows only one station to be in communication with the central station at one time. The central station may be allowed to transmit to several nodes at the same time. Routing messages is very easy because the central station has a unique hardware path to each node. System security is high, since access to the network is controlled by the central station. Another advantage is that priority status can be assigned to selected nodes in the network.

## LAN Protocols

A communication protocol was defined earlier as the set of rules that govern data communications. One of the most important functions of a

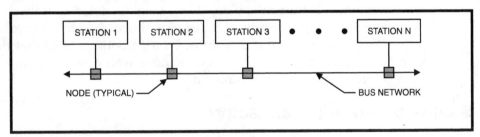

**Figure 9-21. Bus-type LAN.**

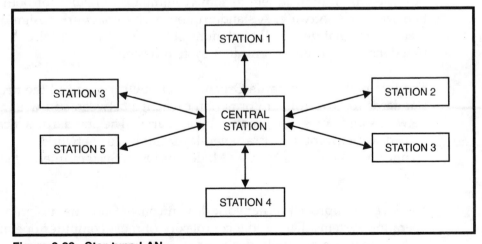

**Figure 9-22. Star type LAN.**

LAN protocol is to govern access to the communications network. Failure to control access would result in chaos any time the traffic on the network rose above a certain minimum level.

The standard LAN protocols are polling, token passing, and carrier sense multiple access/collision detect (CSMA/CD). In *polling*, a master station selects each of the other stations in turn and gives the station permission to communicate for a fixed period of time. The main disadvantage of polling is that each station must remain idle except when selected by the master. This type of protocol is best suited for bus- or star-type LAN topologies, but it is not normally used because stations must remain idle for a large percentage of the time.

The token passing protocol operates by passing a symbolic electronic token from one station to another in the network. Each station may hold the token for a predetermined length of time before passing it. The station that has the token controls the right to transmit to the network.

The carrier sense multiple access/collision detect (CSMA/CD) protocol allows stations to try to communicate whenever they need access. When a station has a message to send, it first listens to determine if anyone else is transmitting ("carrier sense"). When the station detects an idle channel, it transmits data. If two stations detect an idle channel and transmit simultaneously, a collision will occur. In this case, the "collision detect" part of the protocol informs both stations that the communication has failed; then both stations wait a random amount of time before attempting to retransmit.

## Standard Network Architecture

Data communications between computer systems and networks is possible only if they adhere to some common set of rules for both hardware and software. A standard approach to network design or architecture that defines the relations between network services and functions is required in computer system design.

The International Standards Organization (ISO) recognized the need for standards to govern the information exchange between and within networks and across geographical boundaries. The standard, which has gained wide acceptance, is a seven-layer model for network architecture, known as the ISO Model for Open System Interconnection (OSI).

The layered approach to network design comes from the design of operating systems. Due to the complexity of most computer operating systems they are generally designed in sections, each of which contributes a certain function to the operating system. This method of design makes it easier for each section to be refined and redesigned to meet its functional purpose. Finally, all of the layers are integrated to provide a totally functioning operating system.

The same procedure can be used to design a network system. The ISO/OSI model specifies a hierarchy of independent layers that contain modules for performing defined functions. The ISO/OSI model has seven distinct layers, at both the receiver and the transmitter, through which communications must pass, as shown in Figure 9-23.

The function of each layer of the ISO/OSI model is described as follows:

1. The *physical* layer defines the electrical and mechanical requirements of interfacing to a physical medium for transmitting information. When used, this layer must include the software driver for each communications device and the hardware such as interface equipment, the connectors, modems, and the communication cables.
2. The *data link* is used to establish an error-free communications link between network stations over the physical channel. It formats messages for transmission, checks integrity of received data, controls access to and use of the station, and ensures the proper sequence of the transmitted information.
3. The *network control* layer is used to address messages, set up the path between stations, route messages across intervening stations to their destinations, and control the flow of data between stations.

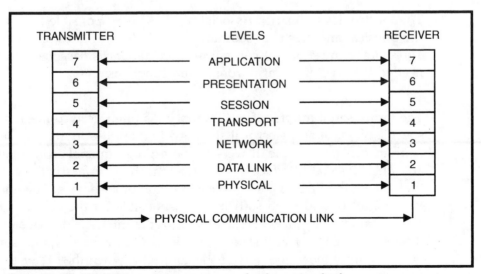

**Figure 9-23. ISO/OSI seven layer communications standard.**

4. The *transport* layer furnishes end-to-end control of a communication once the path has been established, allowing the system to exchange data reliably and sequentially. This layer is normally beyond user control.
5. The *session* layer organizes the dialog of the communication and manages data exchange. This layer is also normally beyond user control, as are layers 6 and 7.
6. The *presentation* layer handles tasks related to data representation and code conversion.
7. The *application* layer handles tasks related to data transfer speed and integrity.

Outside the control of the ISO/OSI are the communications applications processes. Usually, an application process is at the transmitter and another is at the receiver. Computer operations that require a user at a terminal or a piece of software performing instructions are examples of application processes.

## Serial Hardware Standards

Serial hardware standards are required for the physical and electrical compatibility among devices that communicate. Ports and connectors must match in size, shape, and pin configuration. Transmitting and receiving circuits must agree on how signals are generated and interpreted. The so-called "physical layer" communications requirements are handled by the serial interface standards such as EIA232, RS-422, RS-485, and 20 mA current loop.

The Electronics Industries Association (EIA) is responsible for setting the physical and electrical standards in the United States of America. There are many different communication standards because many are needed to serve different purposes and functions.

The term "serial interface" or "interface" is often used to mean a specific physical and electrical standard that applies to a given piece of equipment. For example, if we say that a programmable controller has an "EIA232 serial interface," we mean that the interface is the circuitry that allows the PLC to communicate with another device. This interface circuit and its associated software is responsible for the ISO/OSI Layer 1 physical and electrical characteristics of the interface device as well as the required Layer 7 application functions. On the other hand, some people simply mean the port or the connector when they refer to an EIA232 interface.

## EIA232 Standard

The EIA232 standard is one of the most common serial interfaces in use today. This standard defines a number of physical and electrical characteristics. Most EIA232 connectors have 25 connection pins and have a "D" shape. Each pin, or data line, is assigned a special purpose. In this standard, a logical 1 is a voltage greater than +3 volts dc; logical 0 is a voltage less than -3 volts dc. The format for EIA232 is given in Table 9-1

Data transfer rate up to 25 kilobaud are possible, but cabling between devices is limited to 50 feet. EIA232 is a master-slave serial interface, where there is a primary station in control of a secondary station or stations. The term "unbalanced link" is also used to mean EIA232 communications.

## RS-422 Standard

The RS-422 interface standard, unlike EIA232, creates balanced link communications. It uses two data lines, line A and line B, each line of which transmits and receives. The voltage value of the data signal is determined by the difference in voltage between the two lines sensed by the interface electronic circuitry.

RS-422 is much faster than EIA232 and transmission rates up to 10 megabaud are possible. The interface can support up to 32 drops along the network. Wiring or cabling distances are much longer than with EIA232, up to 4000 feet. This is because the voltage difference, not its specific value, determines whether a signal is 1 or 0. A listing of the important features of the RS-422 standard are given in Table 9-2.

## RS-485 Standard

The RS-485 interface standard is based on impedance sensing rather than voltage sensing. Like the RS-422 interface, RS-485 can be run for distances up to 4000 feet and it can support up to 32 device connections.

| |
|---|
| Voltage sensing:<br>  +12 Vdc = logical 1<br>  −12 Vdc = logical 0<br>  performed at 60 cps |
| 50-foot maximum cable length recommended |
| Point-to-point communications |

**Table 9-1. EIA232 Format.**

| Voltage sensing:<br>  +12 Vdc = logical 1<br>  −12 Vdc = logical 0<br>  performed at 60 cps |
| --- |
| 4000-foot cable length capability |
| 32 device drops |
| Strict wiring guidelines |

**Table 9-2. RS-422C Format.**

The wiring guidelines under RS-485 are very strict. For example, all connections are polarized and cables must be kept a specific minimum distance from power cables and wires. Where power wires cross an RS-485 cable, the crossing must be made at 90° angles. RS-485 standard has greater immunity to noise interference than either EIA232 or RS-422.

A listing of the important features of the RS-485 standard are given in Table 9-3.

## 20 mA Current Loop Standard

The 20 mA current loop standard is very popular in industrial control system applications. The current in the serial communication line is kept at a constant 20 milliamps. Logical 1 and logical 0 are determined by opening and closing the current loop between devices in the network. Current "On" represents logical 1 and current "Off" represents logical 0. Since the current is kept constant at 20 mA for logic 1, the resistance (i.e., length of the communication line) has no effect on the data signal. This means that long network distances are possible. The properties of the 20 mA current loop standard are given in Table 9-4.

| Impedance sensing:<br>  performed at 60 cps |
| --- |
| 4000-foot cable length capability |
| 32 device drops |
| Strict wiring guidelines:<br>  All wires must run at a certain distance from power wires<br>  All power cable crossings at 90 degrees<br>  All wires must make same connections at every device (polarized) |

**Table 9-3. RS-485 Format.**

| Current sensing: |
|:---|
| Current On = logical 1 |
| Current Off = logical 0 |
| 3,000-foot distance capability |
| Convertible to EIA232 or RS-485 |

**Table 9-4. 20 mA Current Loop Format.**

## EXERCISES

9.1 Discuss in detail the function of the three basic components of a communications system.

9.2 Explain the operation of unidirectional, half-duplex, and full duplex communications and give the advantages and disadvantages of each method.

9.3 Compare and discuss serial and parallel data transmission. Give the advantages of each method and some typical applications encountered in computer systems.

9.4 Discuss the three common methods of signal multiplexing encountered in communications systems.

9.5 Generate the ASCII code with odd parity for the message LEVEL HIGH ON TANK 200.

9.6 Explain the four common data transmission error checking methods used in programmable controller communications systems.

9.7 Calculate the LRC/VRC odd parity characters required for the message: MOTOR 1 OFF.

9.8 Explain the difference between asynchronous and synchronous data transmission and give an example of each method.

9.9 Discuss the three common serial data protocols used in communications systems.

9.10 Assemble the bit streams required to send the message MOTOR M-125 ON using 8-bit even parity ASCII code and BISYNC protocol. Assume the header is given by 00000010, which represents station 2, and use a CRC check character at the end of the transmission.

9.11 Discuss the function of each of the seven layers of the ISO/OSI communications standard.

## BIBLIOGRAPHY

1.  *Reference Manual Data Highway/Data Highway Plus Protocol and Command Set*, Allen-Bradley Company, Inc., 1987.

2.  *Digital Industrial Networks Guidebook*, Digital Equipment Corporation, 1988.

3.  Seyer, M. D., *RS-232 Made Easy: Connecting Computers, Printers, Terminals, and Modems*, Prentice-Hall, Inc., 1984.

4.  Stone, H. S., *Microcomputer Interfacing*, Addison-Wesley Publishing Company, 1982.

5.  Thiel, C. A., (Ed.), *IBM Systems Journal Telecommunications*, Volume Eighteen, Number Two, International Business Machine Corporation, 1979.

6.  Svacina, B. *Understanding Device Buses: A Tutorial, Turck Inc.*, 1996.

7.  *Networking: The Competitive Edge*, Digital Equipment Corporation, 1985.

8.  Stacy, A. H., *The Map Book: An Introduction to Industrial Networking*, Industrial Networking Inc. 1987.

9.  Thompson, L. M., *Industrial Data Communications Fundamentals and Applications*, ISA, 1991.

# 10

# System Design and Applications

## Introduction

As a general type of control system, the programmable controller offers a wide variety of system configurations and capabilities. These range from a single machine or process control to an entire industrial plant control and monitoring system. After the decision has been made to use a programmable controller in an control application, the design engineer must complete the system design. This chapter will discuss a detailed system design method and then three typical PLC applications will be presented.

## System Design

The design of a programmable controller-based control system requires a simplified process flow diagram or equipment layout drawing, a process or machine control description, sizing and selection of the PLC equipment, a system specification, system drawings, wiring diagrams, and control programming. Each of these areas will be discussed to give a step-by-step approach to system design.

### Piping and Instrument Drawings

The design of any process control system must start with the piping and instrument drawing (P&ID) or mechanical flow diagrams (MFDs). These drawings show the process and/or mechanical equipment to be controlled and the instrumentation used in the control of the process or machine.

In the process industry, a standard set of symbols is used to prepare the P&IDs and MFDs. The symbols used in these drawings are generally based on American National Standards Institute (ANSI) and Instrument Society of America (ISA) standard, ANSI/ISA-S5.1-1985,

Instrumentation Symbols and Identification. These drawings show the interconnection of equipment and the instrumentation used to control the process. A typical example of a P&ID is shown in Figure 10-1.

In standard P&IDs, the process flow lines, such as process fluid flow and steam flow, are shown as heavy solid lines. The instrumentation signal lines are shown in a way that indicates whether they are pneumatic or electric. A cross-hatched line is used for pneumatic lines, for example, a 3-15 psi signal. The electric signal lines, usually 4-20 mA dc current is normally represented by a dashed line.

A balloon symbol with an enclosed two- to four-letter code is used to represent the instruments associated with the process control loops. For example, the balloon in Figure 10-1 with TT-100 enclosed is a temperature transmitter, and that with TIC-100 enclosed is a temperature-indicating controller. Generally, a number is assigned to each control loop; combining the letter code and number into an instrument tag number labels the specific device in the loop, as illustrated in Figure 10-1.

Special items such as control valves and in-line instruments (for instance, orifice plates) have special symbols, as shown in Figure 10-1. Refer to ANSI/ISA-5.1 for a more detailed discussion of instrument symbols.

The control system design engineer normally reduces the P&ID to a simplified process control diagram that shows only the equipment and

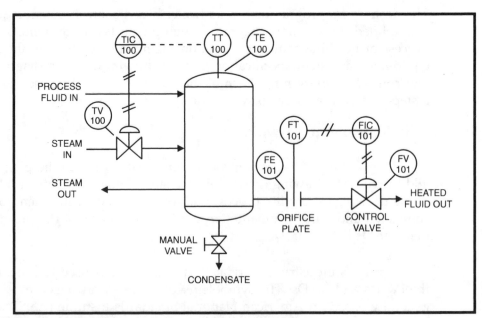

**Figure 10-1. Typical process and instrument diagram.**

instrumentation controlled or measured by the programmable controller. These simplified diagrams are then used to show the status of the process in each step or state to aid in the programming of the system.

## Process Description

The process description is probably the most important step in the design process because it conveys in simple language the purpose and the steps of the process. It is important because in most applications it is the main vehicle of communication between the user and the designer.

## Sizing and Selecting a PLC System

The sizing of a PLC system consists of estimating the number of input/output (I/O) modules required to control the process and the size of PLC memory needed. The selection process also consists of choosing the correct programming language and peripheral equipment required for the control application.

### I/0 Sizing

Many different types of I/O modules may be needed for a given PLC application. Limit switches, push buttons, selector switches, motor controls, solenoids, and pilot lights may require either ac or dc modules of different voltage levels. Solid-state displays and some electronic instrumentation may require +5 V dc logic interface modules. Process instrumentation for measurement of level, flow, temperature, or pressure may require analog-to-digital (A/D) conversion interfaces. Incremental encoders and stepping motors might need special purpose input/output modules.

The interface to these I/O devices can be done with external "black boxes" to condition the signals, which would increase overall equipment cost for a system. Therefore, a PLC should be selected with the correct input/output modules to match the type field devices used in the process.

The number of field devices that can be interfaced to a PLC system is an important consideration in sizing the I/O. Each programmable controller has a maximum number of input and output devices that can be monitored or controlled. Most PLC systems have I/O capacities ranging from a few to 8096. These capacities can be divided into four PLC size categories: micro, small, medium, and large. Micro PLCs have a limited number of built-in I/O circuits normally less than 32 points.

A small PLC system would range from 32 to 256 I/O points, a medium-sized PLC encompasses 256 to 1024 I/O, and a large PLC system has 1024 and higher I/O points. To determine the PLC system size required, simply add up the number of field and control panel devices and compare the total to the above classifications. The system designer must also define the type and number of I/O, because some PLC systems are constrained as to the mixture that can be interfaced to a given I/O system.

If modular I/O systems are being used, input and output point totals can then be used to determine the number and type of I/O modules required. Each type of module can interface a certain number of I/O, such as 4,8,16, or 32. Divide the number of inputs or outputs by the number of I/O points per module and round up to the nearest whole number. This calculation must be performed for each type of I/O module, i.e., discrete, pulse, analog, etc.

After defining the I/O requirements, the designer must also consider spares and future expansions. Most programmable controller users find that 10% to 20% spare capacity is enough for normal system growth.

## Memory Sizing

The amount of memory required for an application is primarily a function of control program complexity and the number of I/O points in the system. The most precise method of determining memory size is to write out the control program and count the number of instructions used in the program. Then multiply this count by the number of words used per instruction. This number can be obtained from the PLC programming manual. Also consult the programming manual on the amount of memory used by executive programs and the processor overhead. The problem with this method is that, in a large system, the system programming sometimes takes months to complete, and the system must be designed and purchased in advance. A shortcut method consists of multiplying the number of I/O points by 10 to obtain a rough estimate the memory required.

An illustration of memory sizing using this simpler method is given in the following example.

---

**EXAMPLE 10-1**

**Problem:** Estimate the memory size required for the PLC application shown in Figure 10-2.

**Solution:** To calculate the memory size, we first need to calculate the number of I/O points in the system in the Figure 10-2.

Remote Area 1: I/O points = 70 + 35 + 6 = 111
Remote Area 2: I/O points = 95 + 50 + 10 = 155
Main Process Area: I/O points = 300 + 156 + 32 + 5 = 493
Total system I/O points = 111 + 155 + 493 = 759

Therefore, memory size = 10 x 759 = 7590 or 8K.

---

## Selecting programming Language

As mentioned earlier, the five basic types of programming languages available in programmable controller systems are ladder diagram, structured text, instruction list, function block diagram, and sequential function chart. The type selected depends on the complexity of the control system and the background of the control system programmer and operators. Most PLCs offer the basic ladder logic instructions plus a combination of the other types of languages. The most common type of language selected is ladder diagram, since this covers the basic ladder logic instructions and some data transfer and manipulations functions.

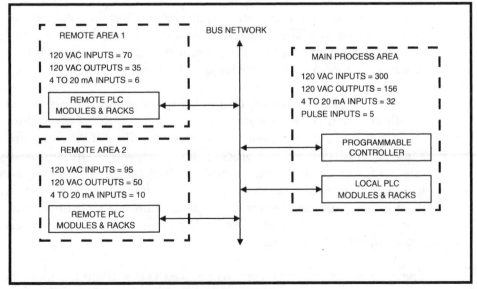

**Figure 10-2. Typical process plant PLC system.**

## Peripheral Requirements

The term "peripheral" refers to the other equipment in the programmable controller system that is not directly connected to field I/O
devices but increases the capabilities of the system. The most common
peripheral is the programming device. This is generally available in
three formats: a compact portable programming device from the PLC
manufacturer, a portable personal computer (PC) with PLC software
installed, or an industrial-type personal computer with programming
software included. The compact portable programmer normally has a
small, limited function key pad and a seven-segment LED display and
can handle only one or two logic rungs at a time. It is normally used
on micro or small PLC systems or for minor field changes to larger
systems. The portable PC-based programming device is normally used
for field start-up and troubleshooting. The desk top personal computer-
based programmer is normally used in a lab or office environment to
perform the development programming on PLC system.

Another common peripheral used in PLC systems is a magnetic tape
storage unit. This unit is used to store the control program on magnetic
tape so that, in the event of a program loss, the backup program can be
reloaded into the controller memory. If a personal computer is used in
the PLC system, the program can also be saved on a floppy disk or a
hard disk for future use.

For hard copy printouts of the control programs, a printer can be
interfaced with the programming device, normally a PC, to obtain a
program listing.

It is important for a complete system design that the peripheral
equipment be available to back up and document the control program
during start-up and field testing, because reprogramming and
documentation can be expensive and time consuming.

Other common peripherals are PROM programmers, process I/O
simulators, and communications modules. The PROM programmers are
used to write and save control programs on the PROM chips used in
some controllers. The I/O process simulators are useful and cost-saving
devices for large and complex systems that can be fully tested before
installation and start-up in the field. The communication peripherals are
used to communicate between the programmable controllers and the
plant or personal computers and other controllers in a system via
communication networks.

The type of operator interface to be used is one the most important
considerations in a PLC system design. There are four main options for

operator interfaces: (1) hard-wired local and main control panels;
(2) graphical interface unit (GUI) software run on a personal computer;
(3) intelligent peripheral devices, such as touch screen operator
interface; and (4) industrial PC with function keys and GUI software.
The system designer might also select a combination of several operator
interface methods to implement a control system, such as hard-wired
local panels and a GUI on a personal computer mounted in a central
control room.

## System Drawing and I/0 Wiring Diagrams

A system drawing is used to give an overall view of the system
hardware (I/O modules, processor, and peripheral equipment) and the
system interface and communication cabling. This drawing is also
useful in identifying all the interface cables by model number. A typical
PLC system drawing is shown in Figure 10-3. The system consists of a
programmable controller, a PC-based programming terminal with GUI
graphics and an installed programming software, a printer, a communications cable, a rack mounted dc power supply, an 8 module I/O
equipment rack, and the associated input/output modules.

A typical discrete output module wiring diagram is given in Figure 10-4.
This drawing shows the wiring of the process control equipment, such
as heaters, pumps, and motors, to an ac output module. The wiring
terminal strip in the programmable controller equipment cabinet is
designated by TB-1, and a field junction box terminal strip is designated
by JB-1 in the example. Field junction boxes are used extensively in
process control applications to simplify field installation and
maintenance of instrumentation. Field wiring is normally indicated on
wiring diagrams by a dashed line, as shown in Figure 10-4. The PLC

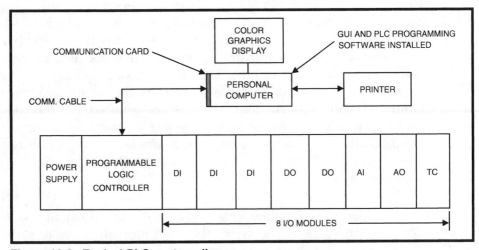

**Figure 10-3. Typical PLC system diagram.**

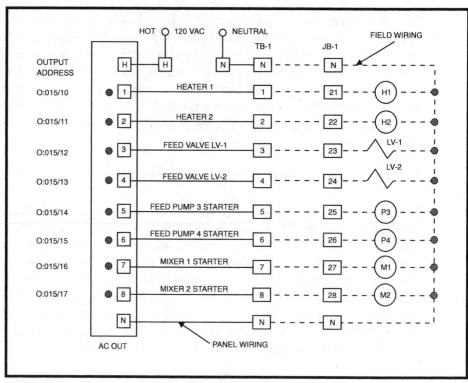

**Figure 10-4. Typical discrete output module.**

output addresses are given on the left-hand side of the wiring diagram
to aid the engineer or technician in the start-up or troubleshooting of
the control system.

## System programming

The programming of a PLC system can be done by the design engineer, plant operations personnel, maintenance, or the control system
integrator. Programming by the system design engineer is the best
choice, since it requires less documentation (flowcharting, process
description, etc.) and less time. The other choice is to have the plant
maintenance or operating personnel perform the system programming.
More up-front documentation and time are required, but this method
results in greater control system acceptance by plant personnel.
Programming by a systems integrator is another choice, but it requires
more documentation and debugging time and more training time for
plant personnel.

The selection of programming language type should usually be left to
plant operations personnel, since they will normally have to maintain
the software after the system is installed.

We are now ready to discuss three PLC-based control system applications.

## Natural Gas Dehydration

The first application to be discussed is using a PLC to control the dehydration process shown in Figure 10-5. The dehydration process removes water from natural gas by using small beads in the process tower to absorb moisture from the gas.

In this process, the differential pressure switches (PSHL-1 and PSHL-2) are used to detect both high and low pressure across the dehydration tanks. The air-operated valves (AOVs) are used to control the gas flow to the two towers shown. A typical diagram of the AOVs is shown in Figure 10-6. An electrically operated solenoid valve is used to supply instrument air at 80 psi to activate the control valves, and the valves have limit switches to indicate if the valves are open or closed.

The limit switches are used by the programmable controller to determine the status or state of the valves. In some applications, only

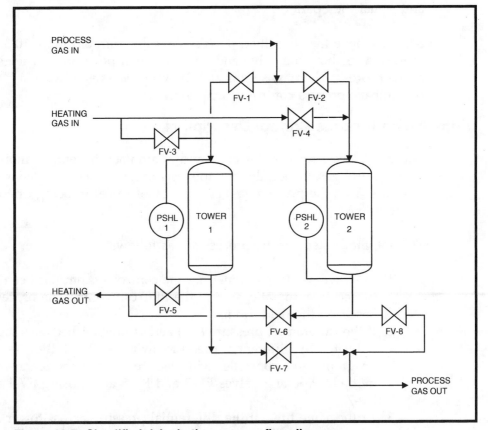

**Figure 10-5. Simplified dehydration process flow diagram.**

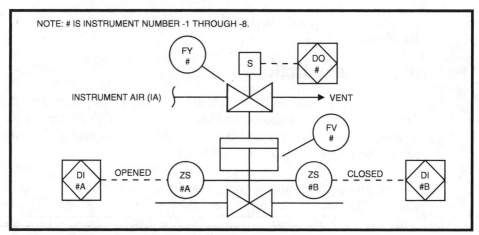

NOTE: # IS INSTRUMENT NUMBER -1 THROUGH -8.

**Figure 10-6.  Detail for air-operated valve.**

one limit switch is used and the programmable controller can assume the opposite state (open or closed) if a single switch is used. However, a more reliable control is obtained if two switches are used. For example, if a valve fails to completely open or close on a given operation, the programmable controller is able to detect this failure and signal the operator.

Since we have the simplified process flow diagram of Figure 10-5, we can use it to show the valve and/or equipment position for each state of the process. The next step in the design process is to write a preliminary process control description.

## Dehydration Process Control Description

The two process towers (towers 1 and 2) are used to remove moisture from natural gas. Generally, one tower is in *service* (i.e., removing moisture from the process gas) and the other tower in being dried out or *regenerated*.

The automatic steps of the process are as follows:

1. First, we need to assume that the control system has been placed in automatic, so that the PLC can control the process based on the field inputs.
2. If the differential pressure (dP) across tower 1 becomes high, as indicated by differential pressure switch (PSH-1), the programmable controller will place tower 1 in the regeneration mode by opening valves FV-3 and FV-5 and closing FV-1 and FV-7.
3. At the same time, if the differential pressure across tower 2 is low, tower 2 will be placed in service by opening the gas valves

FV-2 and FV-8 and control system will close the heating cycle valves FV-4 and FV-6 for tower 2.

4. Later, if the differential pressure across tower 2 becomes high and dP signal across tower 1 returns to low, the control system will place tower 1 in service and tower 2 into regeneration. This cycle will continue until the operator elects to stop the process.

After this preliminary control description is written, the designer is ready to size and select a programmable controller system.

## Sizing and Selecting the PLC System

The first step is to decide on the operator machine interface to use, such as local control panel or personal computer based GUI. In this application, the dehydration process is a part of an entire natural gas processing plant and plant operations personnel have decided to use a personal computer-based GUI for operator interface. The graphics screen for tower 1 in service and tower 2 in regeneration for the dehydration process is shown in Figure 10-7.

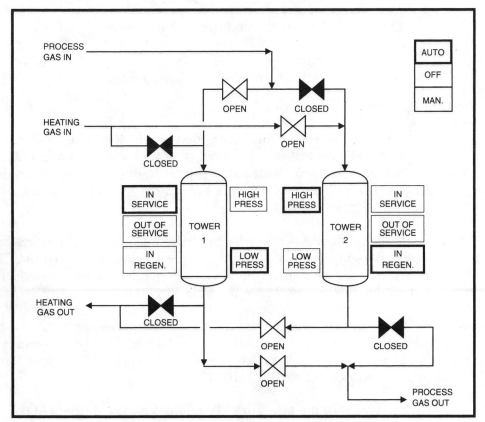

**Figure 10-7. Dehydration GUI graphics display.**

The operator uses the computer mouse to select the mode of operation required for the process. For example, if the pressure across Tank 1 is low (i.e., the low pressure (Low Press) box on the GUI display is highlighted), the operator can click on the In Service (In Serv.) button and place Tank 1 in-service to remove moisture from the natural gas.

The next step in the design process is to calculate the number of input and output modules required This can be done by using the dehydration flow diagram in Figure 10-5 and the detail for the air-operated valves in Figure 10-6. There are eight (8) 120-Vac solenoid valves, so we need eight 120-Vac PLC outputs. There are sixteen (16) limit switches for the open and closed positions of the valves and there are also four differential pressure switches. So, we need 20 discrete input points for this application. We can arbitrarily select 120 Vac for the input signal voltage to save on the types of modules and supply voltages used in the system.

Let us assume that we select Allen-Bradley (A-B) 8-point, 120-Vac I/O modules for use in our application. The number of modules required can be calculated as follows:

Vac input modules
(20 + 20%)/ 8-points/modules = 3 modules

Vac Output modules
(8 + 20%)/ 8-points/modules = 2 modules

## System Drawing

The system drawing for this application will consist of an A-B PLC5/15 programmable controller, a personal computer with GUI and PLC programming software installed, a printer to document the program, and a single I/O rack with 8 slots to hold the five I/O modules, as shown in Figure 10-8.

The system diagram is normally plotted on a "D" size (24 inches by 36 inches) drawing sheet and would include a detailed material list. This list includes the equipment number, material description, manufacturer, and model number, as shown in Table 10-1.

We have selected an Allen-Bradley PLC5-15 for this application and an IBM Pentium personal computer for the GUI and PLC programming software. An 8-slot A-B chassis was chosen to hold the AC input and output modules as listed in Table 10-1.

The placement of the ac input (AI) modules and ac output (AO) modules in the I/O equipment rack is quite arbitrary in this application

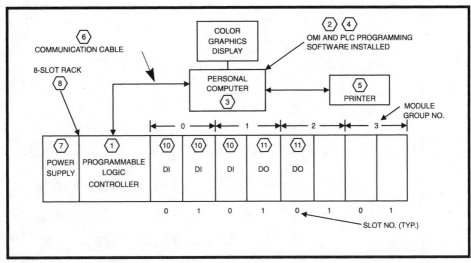

**Figure 10-8. Dehydration PLC system drawing.**

since we have selected only ac modules. If the system has a mix of ac and analog modules, the ac modules are normally integrated from low level signal modules such as analog I/O (4 to 20 mA dc), 5 Vdc inputs, and millivolt (mV) inputs to avoid electrical interference.

## I/O Wiring Diagrams

The I/O wiring diagram for the first ac output module is shown in Figure 10-9. This drawing shows the wiring of the system solenoid valves to the ac output module.

In our example application, a single ac line is used where the ac hot is normally designated by L1 and ac neutral has wire number L2. However, in larger systems, more than one ac line might be used, depending on loading requirements. The design engineer must also consider maintenance of the system, so that in our example application

| Equip. No. | Material Description | Manufacturer | Model No. |
|---|---|---|---|
| 1 | Programmable Controller | Allen-Bradley | 1785-LT |
| 2 | PLC programming software | Allen-Bradley | 6200 |
| 3 | Personal Computer | IBM | Pentium |
| 4 | GUI Software | Rockwell | RS-View |
| 5 | Graphics Printer | Hewlett Packard | Laserjet III |
| 6 | Communication Cable | Allen-Bradley | 1784-CP |
| 7 | PLC5 rack power supply | Allen-Bradley | 1775-P1 |
| 8 | 8-slot PLC chassis | Allen-Bradley | 1771-A2B |
| 9 | 120 Vac input module | Allen-Bradley | 1771-IA |
| 10 | 120 Vac output module | Allen-Bradley | 1771-OA |

**Table 10-1. Dehydration Control System Material List.**

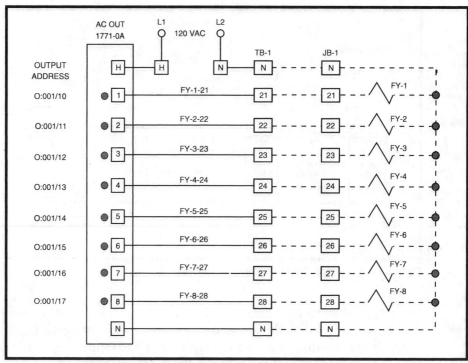

**Figure 10-9. Wiring diagram for valves on dehydration system.**

we might connect the wiring for tower 1 to one ac circuit and the wiring to the other tower to a second ac circuit. In this case, one tower could be taken out of service and the other tower could be kept on line during maintenance of the second tower. Furthermore, the wiring for the control panel might be placed on a third ac power circuit.

The wiring of the ac input modules is performed in a similar manner, except that the field inputs are normally drawn on the left side of the input module as shown in Figure 10-10. This input wiring diagram shows the connection of the high and low pressure switches for towers 1 and 2 to the first AC input module. As shown in the system drawing, the first ac input module is in rack 00, module group 0, slot 0, so that the Allen-Bradley input addresses are between bit I:000/00 and bit I:000/07, as shown on the right-hand side of the module input drawing.

The input wiring diagram in Figure 10-11 shows the connection of the valve limit switches for control valves FV-1 through FV-4. As shown in the system drawing, this ac input module is in rack 00, module group 0, slot 1, so that the input addresses are between bit 111/10 and bit 111/17, as shown on the right-hand side of the module input drawing.

The wiring diagram for the limit switches on the remaining four control valves (i.e., FV-5, FV-6, FV-7, and FV-8) is shown in Figure 10-12. This

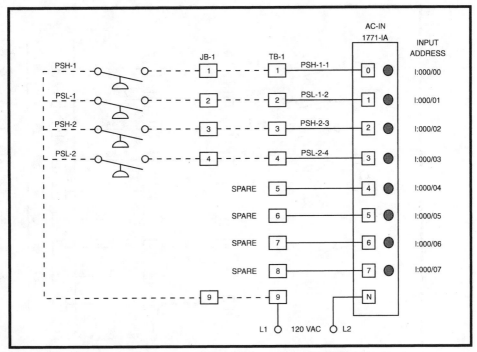

Figure 10-10.  Wiring diagram for differential pressure switches.

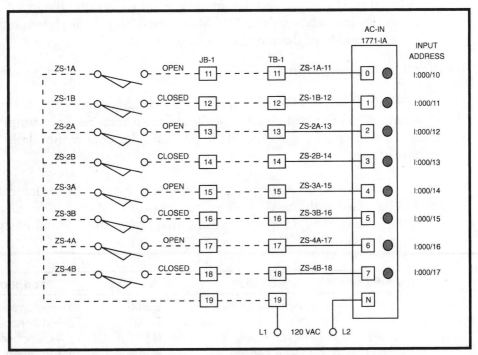

Figure 10-11.  Wiring diagram for valve position switches (ZS-1 through 4).

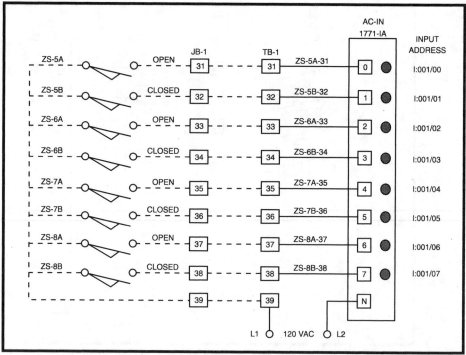

**Figure 10-12. Wiring diagram for valve position switches (ZS-5 through 8).**

ac input module is in rack 00, module group 1, so that the input addresses are between bit I:001/00 and bit I:001/07, as shown on the right-hand side of the module input drawing.

## Application programming

In a small system like the dehydration application, basic ladder programming is the best choice since the system requires only simple on/off control and there are no involved analog or data manipulations anticipated.

The first step in the logic programming is to make a list of internal bits that interface with the GUI. The next step is to list the I/O points in the system using the I/O module wiring diagrams. The internal bits that interface with the GUI are listed in Table 10-2.

| Bit Address | Description | Bit Address | Description |
|---|---|---|---|
| B3/00 | Automatic mode | B3/05 | T1—In regeneration |
| B3/01 | Off | B3/06 | T2—In service |
| B3/02 | Manual mode | B3/07 | T2—Out of service |
| B3/03 | T1—In service | B3/08 | T1—In regeneration |
| B3/04 | T1—Out of service | | |

**Table 10-2. GUI Bit Assignments.**

These bit assignment tables are an aid in programming because they consolidate the I/O information in a single table for easy reference. In most PLC programming software packages, these lists can be entered directly into the control program.

The operation of this ladder program is relatively simple. When the operator clicks on the AUTO button shown in the upper righthand corner of Figure 10-7, this set internal bit B3/00 to logic 1 and places to control system into the automatic mode. The GUI software is programmed to deactivate the OFF and manual (MAN.) buttons on the screen, if the AUTO mode is selected by the operator.

We can write the ladder logic for the automatic mode of the process, as shown in Figure 10-13. Tower 1 is placed in service, if the differential pressure in the tower is low, the AUTO function (internal bit B3/00) is true, and the tower 1 differential pressure is not high (bit I:000/00). The output for "tower 1 in service" is sealed in with its own control bit B3/3. Tower 1 will stay "in service" until the differential pressure in the tower reaches a high level, as measured by a high differential pressure (PSH-1) across the tower. If the dP switch PSH-1 is closed, then input bit I:000/00 is set to true and its normally closed contacts are opened. So the "tower 1 in service" output coil bit B3/03 is turned OFF.

Tower 1 is placed into regeneration by the control system to remove moisture that has built up during the service cycle using the ladder logic software shown in Figure 10-14.

We are now ready to control the flow valves on tower 1. The valves are designed to fail closed (FC), so if plant air or electric power is lost, the valve will close. Therefore, the solenoid valves must be energized to

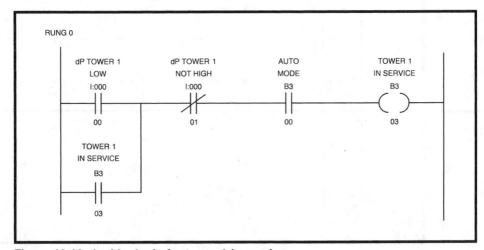

**Figure 10-13. Ladder logic for tower 1 in service.**

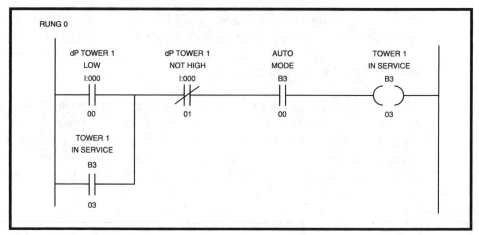

**Figure 10-14. Ladder logic for tower 1 in regeneration.**

open the valve. The ladder logic program in Figure 10-15 shows the control logic for solenoid valves FY-1 through FY-8.

The same design procedure can be used to write the control software for tower 2. To this point, we have not used the open and closed limit switches in our system program. The valve limit switches are used to

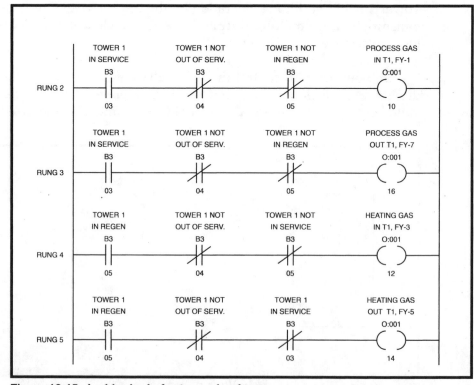

**Figure 10-15. Ladder logic for tower 1 valves.**

animate process graphics valve symbols during each step in the process. If a valve is closed, the valve symbol will be filled in by the GUI software based on the information sent to the PC from the PLC.

## Two-Stage Alternating Pump Application

In this application, we will develop the ladder logic program to alternate pumps in processes like the emptying of wells, reservoirs, and tanks where the rate of flow into the tank is not constant. In this type of application, two smaller pumps are frequently used instead of one large one to reduce cost.

Alternating pump operation (pump 1 as the primary, then pump 2 as the primary) reduces the maintenance required on the individual pumps and provides more reliable operation. In this application, the secondary or "standby" pump is available if the rate of water entering the vessel is more than the first pump can handle. If this situation occurs, the second pump will also turn On and assist the primary pump. The triggers for these events could be analog signals, or simple discrete inputs (level switches, etc.).

The first step in the design of the PLC-based control system is to draw the piping and instrument diagram as shown in Figure 10-16.

The next step in the design procedure is to select a PLC and I/O types. Since there are only five discrete inputs and two discrete outputs, a Micro PLC is sufficient for this application. We can arbitrarily select an A-B Micro 1000 PLC with ten discrete inputs and six discrete outputs.

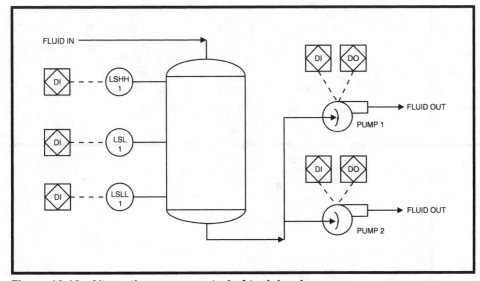

**Figure 10-16. Alternating pump control of tank level.**

We can now complete the design of input/output wiring diagram as shown in Figure 10-17.

The ten inputs (addresses I/0 through I/9) are shown on the left side of the micro PLC and the five AC outputs (addresses O/0 through O/9) are drawn on the right side of the unit. In this diagram, we have connected the three level switches—LSHH-1, LSL-1, and LSLL-1—to the first three inputs on the micro-PLC I/0, I/1, and I/2 respectfully. We have connected the two auxiliary (aux) contacts from the pump motor starters (M1 and M2) to inputs I/4 and I/5. The pump motor starter relays are connected to the first two AC outputs on the micro PLC.

The final design step is to write the control logic for this application. The ladder logic used in this application consists of only four rungs. Rungs 0 and 1 form a alternating circuit, so that each time the fluid in the tank reaches the low-low level switch (LSLL-1) and sets bit I/2, the alternator bit in rung 1, B3/2 changes state. The status of this bit determines which pump will be the first to turn on. The one-shot rising (OSR) instruction in rung 0 is a specialized instruction that is only energized for one processor scan. This causes internal bit B3/1 to be

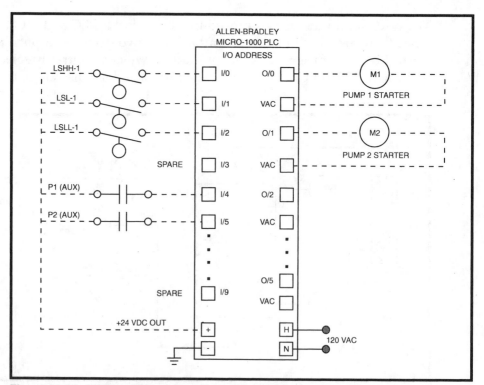

**Figure 10-17. Wiring diagram for alternating pump control.**

energized for one processor scan, if the low-low level switch (bit I/2) is closed as shown in Figure 10-18.

Rung 2 controls the operation of pump 1, using output bit O/0 on the micro PLC. If the low-low level switch is closed it sets bit I/2 and the alternator internal bit B3/2 is turned off, and the level in the tank has reached the low level switch (LSL-1), this pump will be the first one energized. If the alternator bit B3/2 is On, pump l will be the second pump energized as shown in Figure 10-19.

Rung 3 controls the operation of pump 2 using output bit 0/1. If the low-low level switch is On (bit I/2 is set), and the alternator bit B3/2 is

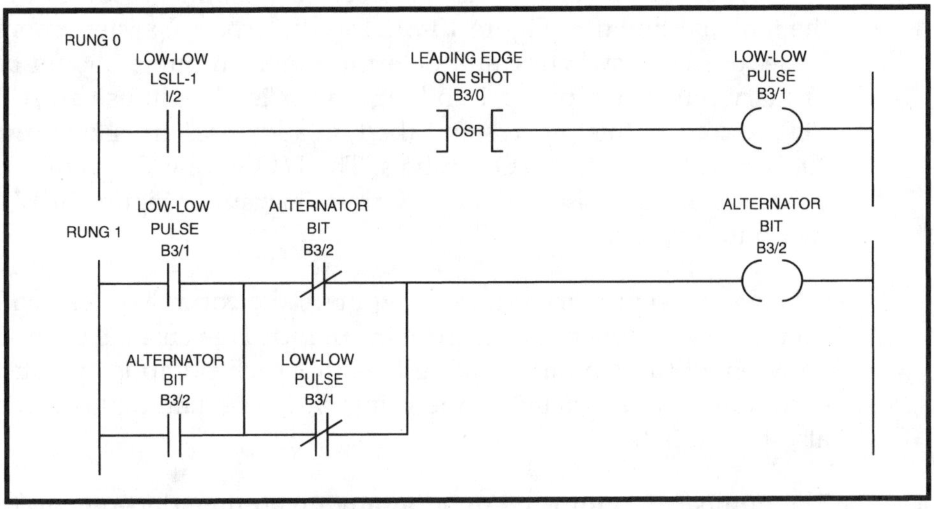

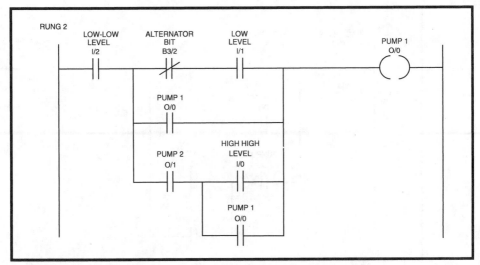

Figure 10-18. Ladder logic for alternating pump control (rungs 0 & 1).

Figure 10-19. Ladder logic for alternating pump control (rung 2).

ON, and the level in the tank has reached the low level switch (i.e., Bit I/1 is set), this pump will be the first one energized. If B3:3/2 is OFF, pump 1 will be the second pump to energized as shown in Figure 10-20.

## Three Station Alternator

This application is similar in function to the two pump example in the last section, except that we have added another pump to alternate three rather than two pumps. For ease of description, we will discuss three pumps that empty the tank. The control system needs to be able to rotate the pump that turns On first each time a request is made, and also to bring other pumps on-line as demand increases.

A series of four level switches are used to monitor the level of fluid in the tank as shown in Figure 10-21. The PLC control system monitors these four level switches, and determines which pump should be the primary pump, lag pump 1 and lag pump 2. We will use an A-B micro PLC to control this process and the I/O addresses are shown in Figure 10-21 next to the PLC I/O symbols. The I/O wiring diagram for this process would be similar to the wiring diagram of Figure 10-17 for the two pump system.

Whenever the primary pump is required, the control system will then rotate the assignment of the primary pump. This ensures even wear between all three pumps and verifies that each pump is operational. As each pump is designated as the primary, the remaining lag pumps will also be rotated.

The number of pumps required at any given time depends on the status of the four tank level switches. For example, if the low low level

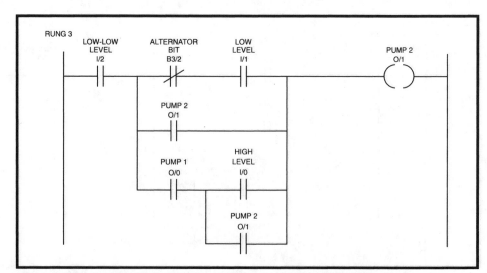

**Figure 10-20. Ladder logic for alternating pump control (rung 3).**

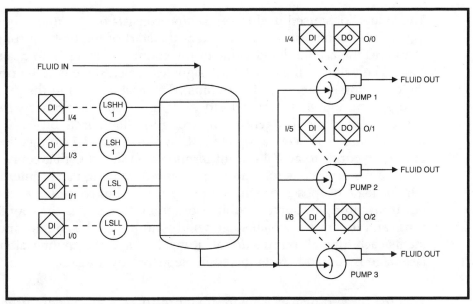

**Figure 10-21. Using three alternating pumps to control tank level.**

switch (LSLL-1) is off, indicating the fluid level in the tank is below the switch, no pumps are required. Table 10-3 lists the pump requirements based on all the possible states of the four level switches on the process tank.

A breakdown of priorities for each pump at any given time is given in Table 10-4. The sequence for the running of each pump is called a staging. There are three pumps, and therefore three stages that operate as shown in the table. For example, stage 1 is defined as follows: Pump 1 is the primary pump, pump 2 is the first lag pump, and pump 3 is the second lag pump.

| Input Device Status | Pump Requirements |
| --- | --- |
| LSLL Off | All pumps Off |
| LSLL On | All pumps Off |
| LSLL and LSL On | Primary pump On |
| LSLL, LSL, and LSH On | Primary and lag 1 pumps On |
| All level switches On | All pumps On |

**Table 10-3. Pump Requirements for Three Station Alternator.**

| Stage | Pump 1 | Pump 2 | Pump 3 |
| --- | --- | --- | --- |
| 1 | Primary | Lag 1 | Lag 2 |
| 2 | Lag 2 | Primary | Lag 1 |
| 3 | Lag 1 | Lag 2 | Primary |

**Table 10-4. Pump Operational Stages.**

The ladder logic used in this application consists of 15 rungs. The EQUAL TO comparison instruction at the start of the first ten rungs compares the accumulated value of the counter in rung 13 to a constant. The value of the constant designates which stage is to be run (i.e., the operating sequence of the pumps). The first 9 rungs in the program (rungs 0 through 8) set the priority assignment for the primary, Lag #1, and Lag #2 pumps. The tenth rung (rung 9) sets internal bit B3/9 when the final stage has been completed. Rungs 10, 11, and 12 rungs link the preceding rungs to actual output terminals on the PLC that connect to the pump starter coils for pumps 1, 2, and 3. Making use of internal bits for logic purposes provides an easy method of controlling an output from multiple sources within a program. Rung 13 is the counter rung, and it controls which stage will be run next. The counter increments each time all pumps are off Rung 14 resets the counter after the last stage is run and starts the entire sequence over again.

The first three rungs are shown in Figure 10-22, and they are used to set the internal bits for the three pumps in stage 1.

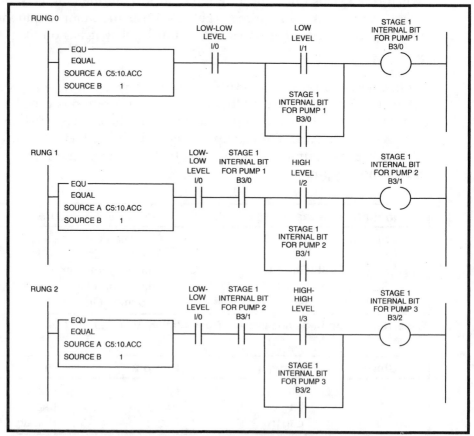

**Figure 10-22. Alternating pump control, stage 1 (rungs 0, 1, & 2).**

The next three rungs (rungs 3, 4, and 5) are shown in Figure 10-23, and they are used to set the internal bits for the three pumps in stage 2.

The next three rungs (rungs 6, 7, and 8) are shown in Figure 10-24, and they are used to set the internal bits for the three pumps in stage 3.

Rung 9 is shown in Figure 10-25. This rungs set the cycle complete bit B3/9 when the accumulated value of the sequence counter reaches 3 and the stage 3 internal bit (B3/6) for pump 1 is set. This bit (B3/6) is set when the pump sequencing cycle detailed in Table 10-4 is complete.

The next three rungs (rungs 10, 11, and 12) use the internal staging bits generated in the first nine rungs to turn on the three pumps in the proper order listed in Table 10-4.

The ladder logic shown in Figure 10-26 is used to turn on pump 1 (control bit O/0) if stage 1 internal bit for pump 1 is true, or if stage 2 internal bit for pump 1 is true, or if stage 3 internal bit for pump 1 is true.

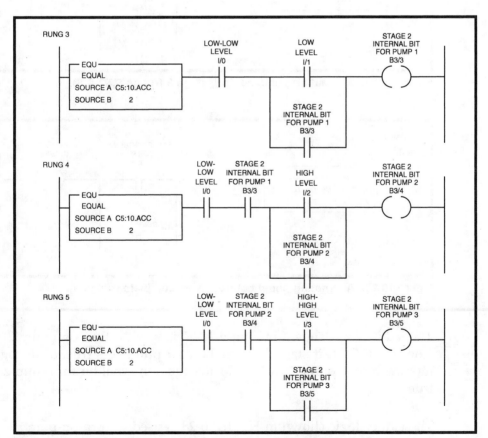

**Figure 10-23. Alternating pump control, stage 2 (rungs 3, 4 & 5).**

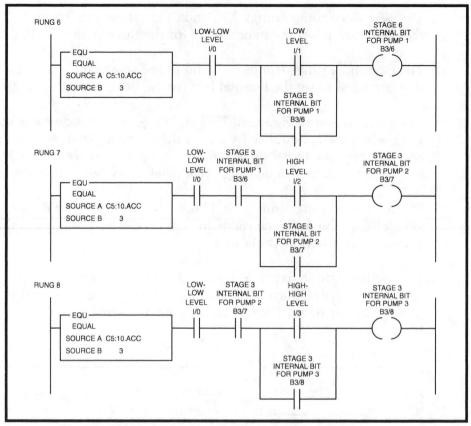

**Figure 10-24. Alternating pump control, stage 3 (rungs 6, 7, & 8).**

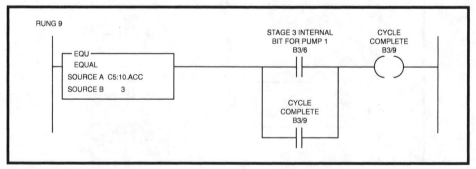

**Figure 10-25. Alternating pump control, cycle complete (rungs 9).**

The ladder logic shown in Figure 10-27 is used to turn on pump 2 (control bit O/1) if stage 1 internal bit for pump 2 is true, or if stage 2 internal bit for pump 2 is true, or if stage 3 internal bit for pump 2 is true.

The ladder logic shown in Figure 10-28 is used to turn on pump 3 (control bit O/2) if stage 1 internal bit for pump 3 is true, or if stage 2

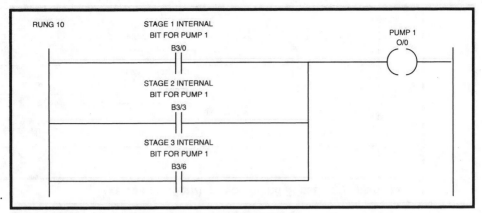

**Figure 10-26. Alternating pump control for pump 1.**

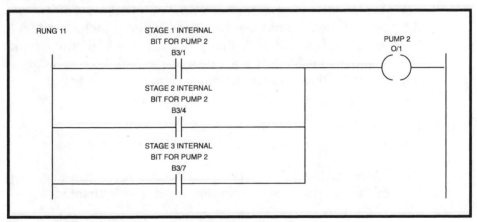

**Figure 10-27. Alternating pump control for pump 2.**

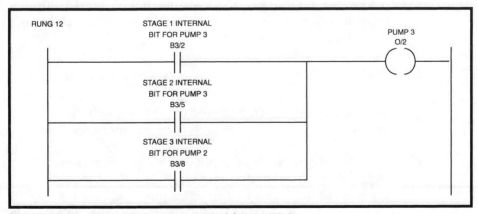

**Figure 10-28. Alternating pump control for pump 3.**

internal bit for pump 3 is true, or if stage 3 internal bit for pump 3 is true.

The final two rungs of ladder logic are shown in Figure 10-29. Rung 13 is the pump sequencer counter and it controls which stage is turned on

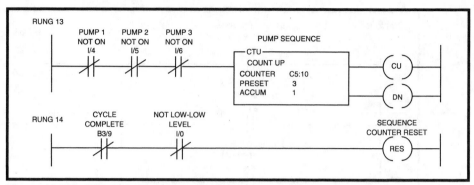

**Figure 10-29. Alternating pump control (rungs 13 and 14).**

next. The counter increments each time all pumps are off by reviewing the status of the auxiliary contacts on the pump motor starters (i.e., input bits I/3, I/4, and I/5). Rung 14 resets the sequence counter if the cycle complete bit B3/9 is not true and the tank level low-low switch is not set (bit I/O) and starts the entire sequence over again.

## EXERCISES

10.1 Write the ladder logic program needed to control tower 2 of the dehydration system discussed in this chapter.

10.2 Draw a revised piping and instrument drawing showing the process equipment and instrumentation needed to provide a third process tower to allow for maintenance of the dehydration process.

10.3 Revise the dehydration GUI graphics to include tower 3.

10.4 Write a process description for the revised dehydration system with a third tower added for maintenance.

10.5 Calculate the number of input and output modules required on the dehydration system if a third process tower is added. Assume that all inputs and outputs are 120 Vac.

10.6 Redesign and redraw the I/O module wiring diagrams if a third process tower is added to the dehydration system. Use four separate ac supply lines (L1, L2, L3, and L4), one line for each process tower, and a fourth on the control panel.

## BIBLIOGRAPHY

1.  Jones, C. T., and Bryan, L. A., *Programmable Controllers: Concepts and Applications,* International Programmable Controls, Inc., 1983.
2.  Hughes, T. A., *Basics of Measurement and Control,* Instrument Society of America, Second Edition, 1995.
3.  *Processor Manual PLC-5 Family Programmable Controllers,* Allen-Bradley Co., Inc., 1987.
4.  *Programming and Operations Manual, PLC-2/30 Programmable Controllers,* Allen-Bradley Co., Inc., 1988.
5.  *Allen-Bradley Industrial Computer and Communications Group Product Guide,* Allen-Bradley Co., Inc., 1987.
6.  *Allen-Bradley Programmable Controller Products,* Allen-Bradley Co., Inc., 1987.
7.  Rockis, G., and Mazur, G., *Electrical Motor Controls, Automated Industrial Systems,* American Technical Publishers, Inc., Second Edition, 1987.
8.  *Micro Mentor: Understanding and Applying Micro Programmable Controllers,* Allen-Bradley Company, Inc., 1995.

# 11

# Installation, Maintenance, and Troubleshooting

## Introduction

To complete the discussion of programmable controller system concepts and design principles, we need to cover panel design, equipment installation and layout, system operational testing, maintenance, and troubleshooting procedures. The reliability and maintainability of a control system is to a large part a function of proper system design that considers the maintenance aspects of a system.

The final area to be discussed will be the troubleshooting methods used to find and correct PLC system problems.

## Control Panel Design

The main feature that separates PLCs from other types of computers is that programmable controllers are designed to be installed in harsh industrial environments. They are, in some cases, simply installed on metal sheet (subpanel) on the production floor. However, in most cases, programmable controller system components are installed in a metal enclosure or a control panel to protect against atmospheric contaminants such as dust, moisture, oils, and other corrosive airborne substances. These metal enclosures also reduce the effects of electromagnetic radiation generated by electrical or welding equipment.

The panel or enclosure design requires a panel layout design, considerations for heating and maintenance, wiring layout and duct design, power distribution design, and normally the writing of a panel specification.

### Control Panel Layout

The panel size depends on the amount of equipment to be installed in the enclosure and whether front panel controls and instruments are required on the system. If front panel controls are required, the size and shape of the panel are mainly controlled by the best layout design of these front panel instruments. To assure correct and easy operation of the system, the indicators and recorders are placed at normal eye level and hand switches are placed below the indicators.

The metal enclosure should conform to industrial standards, such as the National Electrical Manufacturers Association (NEMA) standards. The NEMA standard covers the design of industrial enclosures for different industrial environments.

The equipment layout inside the panel should follow the recommendations contained in the programmable controller installation manual. In this manual, the manufacturer generally lists the minimum spacing allowed between I/O racks, other equipment, and the processor. This spacing generally takes into consideration equipment heating, electrical noise, and safety factors. Figure 11-1 shows a typical minimum equipment spacing of 6" for a programmable controller system per a PLC manufacturer's recommendation.

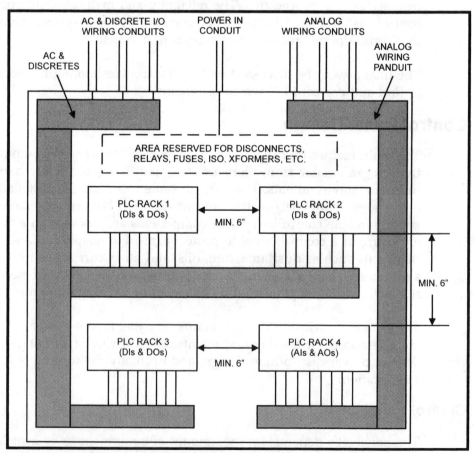

**Figure 11-1. PLC equipment layout diagram.**

## Heating Considerations

To allow for effective convection cooling, most manufacturers recommend that all system components be mounted in a position that allows for maximum air flow in the enclosure. Since the power supplies generate the most heat, they should not be mounted directly underneath another system component. Generally, the main power supply is mounted near the top of the enclosure, but some PLC manufacturers use individual power supplies on each I/O rack and an internal power supply for the processor. In Figure 11-1, notice the 6-inch horizontal gap between the I/O racks to allow for adequate cooling air flow.

The temperature inside the control panel or enclosure must not exceed the maximum operating temperature listed in the manufacturer's installation manual (typically 120°F). If the temperature limit cannot be maintained by convection cooling, a fan or blower must be installed to help dissipate the heat. The fan or fans used will generally be equipped with filters to prevent dust, dirt, and other airborne contaminants from entering the enclosure and affecting system components.

## Enclosure Standards

All enclosures installed in industrial applications must meet the NEMA standard in publication number 250-1979. The following descriptions are excerpts from this NEMA standard and the enclosure types are rated for "Nonclassified Locations" and "Classified Locations" (i.e. explosive locations).

### Nonclassified Location Enclosures Types

*Type 1 Enclosures*   Type 1 enclosures are intended for indoor use primarily to provide a degree of protection against contact with the enclosed equipment in locations where unusual service conditions do not exist. The enclosures shall meet the rod entry and rust-resistance design tests.

*Type 2 Enclosures*   Type 2 enclosures are intended for indoor use primarily to provide a degree of protection against limited amounts of falling water and dirt. These enclosures shall meet rod entry, drip, and rust-resistant design tests. They are not intended to provide protection against conditions such as internal condensation or internal icing.

*Type 3 Enclosures*   Type 3 enclosures are intended for outdoor use primarily to provide a degree of protection against windblown dust, rain, sleet, and external ice formation. They shall meet rain, external icing, dust, and rust-resistance design tests. They are not intended to provide protection against conditions such as internal condensation or internal icing.

*Type 3R Enclosures*   Type 3R enclosures are intended for outdoor use primarily to provide a degree of protection against falling rain, sleet, and external ice formation. They shall meet rod entry, rain, external icing, and rust-resistance design tests. They are not intended to provide protection against conditions such as dust, internal condensation, or internal icing.

*Type 4 Enclosures*   Type 4 enclosures are intended for indoor or outdoor use primarily to provide a degree of protection against windblown dust and rain, splashing water, and hose-directed water. They shall meet hose-down, dust, external icing, and rust-resistance design tests. They are not intended to provide protection against conditions such as internal condensation or internal icing.

*Type 4X Enclosures*   Type 4X enclosures are intended for indoor or outdoor use primarily to provide a degree of protection against corrosion, windblown dust and rain, splashing water, and hose-directed water. They shall meet the hose-down, dust, external icing, and corrosion-resistance design tests. They are not intended to provide protection against conditions such as internal condensation or internal icing.

*Type 5 Enclosures*   Type 5 enclosures are intended for indoor use primarily to provide a degree of protection against dust and falling dirt. They shall meet the dust and rust-resistance design tests. They are not intended to provide protection against conditions such as internal condensation.

*Type 6 Enclosures*   Type 6 enclosures are intended for indoor or outdoor use primarily to provide a degree of protection against the entry of water during occasional temporary submersion at a limited depth. They shall meet submersion, external icing, and rust-resistance design tests. They are not intended to provide protection against conditions such as internal condensation, internal icing, or corrosive environments.

*Type 6P Enclosures*   Type 6P enclosures are intended for indoor or outdoor use primarily to provide a degree of protection against the entry of water during prolonged submersion at a limited depth. They shall meet air pressure, external icing, and corrosion-resistance design tests. They are not intended to provide protection against conditions such as internal condensation or internal icing.

*Type 11 Enclosures*   Type 11 enclosures are intended for indoor use primarily to provide, by oil immersion, a degree of protection to enclosed equipment against the corrosive effects of liquids and gases. They shall meet drip and corrosion-resistance design tests. They are not intended to provide protection against conditions such as internal condensation or internal icing.

*Type 12 Enclosures*   Type 12 enclosures are intended for indoor use primarily to provide a degree of protection against dust, falling dirt, and dripping noncorrosive liquids. They shall meet drip, dust, and rust-resistance tests. They are not intended to provide protection against conditions such as internal condensation.

*Type 12K Enclosures*   Type 12K enclosure with knockouts are intended for indoor use primarily to provide a degree of protection against dust, falling dirt, and dripping noncorrosive liquids other than at knockouts. They shall meet drip, dust, and rust-resistance design tests. Knockouts are provided in the top and/or bottom walls only. After installation, the knockout areas shall meet the environmental characteristics listed above. They are not intended to provide protection against conditions such as internal condensation.

*Type 13 Enclosures*   Type 13 enclosures are intended for indoor use primarily to provide a degree of protection against dust, spraying of water, oil, and noncorrosive coolant. They shall meet oil exclusion and rust-resistance design tests. They are not intended to provide protection against conditions such as internal condensation.

## Classified Location Enclosures

*Type 7 Enclosures*   Type 7 enclosures are for indoor use in locations classified as Class I Groups A, B, C, or D, as defined in the *National Electrical Code.*

*Type 8 Enclosures*   Type 8 enclosures are for indoor or outdoor use in locations classified as Class II, Groups A, B, C, or D, as defined in the *National Electrical Code.*

*Type 9 Enclosures*   Type 9 enclosures are intended for indoor use in locations classified as Class II, Groups E, F, or G, as defined in the *National Electrical Code.*

*Type 10 Enclosures (MSHA)*   Type 10 enclosures shall be capable of meeting the requirements of the Mine Safety and Health Administration, 30 C.F.R., Part 18 (1978).

## Maintenance Features

The system designer must include certain features in the design of the enclosure to reduce maintenance time and cost. One important consideration is the accessibility of equipment components and terminal connections. For example, the processor should be placed at a normal working level for ease of operation or maintenance. If the processor and the system power supply are contained in a single unit, it should be placed near the top of the enclosure. However, if there is adequate space in the enclosure, it should be placed to improve operation or maintenance.

Another maintenance feature is that the control panel should have an ac power outlet strip so maintenance can plug in test equipment and a portable light if needed. If the panel is large, interior lighting should be installed to aid maintenance and operations personnel in trouble-shooting. Both the ac power strip and the interior lighting should be on a separate ac circuit from the other system components.

In some applications, a gasketed Plexiglas™ window is used to allow for viewing of the processor status lights and/or I/O status indicators by operations and maintenance personnel. In most applications, the operator will check the status of process and I/O lights during each step in a process to make sure there are no abnormal operating conditions.

## Panel Duct and Wiring Design

The wiring duct layout is determined by the types of signals used in the system design. It also depends on the placement of the I/O modules in the racks. The main consideration is to reduce the electrical noise caused by crosstalk between I/O signal lines.

All ac power wiring should be kept separated from low level dc wires. If dc lines must cross ac signal or power lines, it should be done at right angles only. This routing practice minimizes the possibility of electrical interference.

## Power Distribution Design

Most programmable controller manufacturers recommend that a power isolation transformer be installed between the ac power source and the PLC equipment to provide for signal isolation from other equipment in the process area. A typical power distribution drawing is shown in

Figure 11-2. In this application, the incoming ac lines (L1, L2, and L3) are 120 V ac, used for power devices, such as motors, heaters, or pump starters, in the system. There are three fuses on the incoming lines to protect against any overcurrent condition. In the drawing, the 120 V ac power lines L1 and L2 are connected to the primary of the step-down transformer. This transformer steps down or reduces the ac voltage to 120 volts and provides isolation for the programmable controller components from the outside electrical equipment.

The ac distribution circuit contains a "master control relay," labeled as MCR in the circuit diagram. This relay is used to stop the operation of the programmable controller or the controlled machine or process when any emergency stop (E-stop) pushbutton is depressed by an operator. Any number of E-stop switches can be used in the ac distribution circuit to improve system safety.

It is also recommended that emergency stop circuits be designed into the system for every machine being directly controlled by the programmable controller. These circuits should be hard wired and totally independent of the programmable controller to provide for maximum safety in the control system.

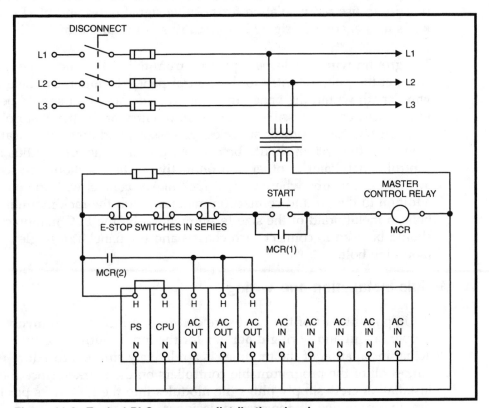

**Figure 11-2. Typical PLC ac power distribution drawing.**

Programmable controllers are very reliable devices, but failure of the central processor can cause dangerous and erratic behavior of the control system. Therefore, the operator must be able to quickly and safely turn off process equipment and machinery by using E-stop switches that are placed in locations easily accessible to the operator.

## Grounding Considerations

The reliability of any electrical control system is highly dependent on the proper design of system grounding. Correct electrical grounding design is also required to guarantee a safe electrical installation. When designing and installing any electrical equipment, the designer should read and understand Section 250 of the National Electrical Code (NEC). This section provides information on the wire color code, size, and type of conductor required as well as connection methods required for safe grounding of electrical equipment.

The code states that a ground must be permanent (i.e., no solder connections), continuous, and able to safely conduct the current in the system with minimal resistance. The ac hot wire must have black-colored insulation, the ac neutral wire insulation must be white, and the ground conductors must use green-colored insulation. It is common practice to use red insulation for positive signal wires and black-colored wires for negative dc signal lines, but it is not required by NEC.

The ground wire should be separated from the ac hot and ac neutral wires at the point of entry to the control panel. To minimize the ground wire length within the panel, the ground reference point should be located as close as possible to the point of entry of the panel supply line. All I/O racks, power supplies, processors, and other electrical devices in the system should be connected to a single ground bus in the control panel. Paint or other nonconductive materials should be scraped away from the area where an I/O rack makes contact with the panel. In addition to the ground connection made through the rack mounting bolts, a metal braid of the size recommended by the PLC manufacturer should be used to connect each chassis and the panel at a single mounting bolt.

## I/O Module Installation and Wiring

An important part of the hardware design of a PLC system involves the proper layout and wiring diagrams for the input/output module. The actual installation of the modules is a relatively simple procedure with almost all of the programmable controllers on the market, since the installation crew simply plugs the modules into the I/O racks per the drawing provided by the system designer.

However, in some cases, intelligent modules (such as thermocouple, analog, communications, etc.) might require switch settings to properly configure the module. For example, a thermocouple (T/C) module might be designed to accept the various T/C types, such as J, K, T, S, etc., and the user sets toggle switches in the module to accept the T/C being used in the process. The design engineer must document the switch setting on the I/O drawings or in the installation instructions to guarantee that the modules are properly installed and configured.

Another important procedure required for installation of a programmable controller system is the selection of the I/O chassis number for each I/O rack. Many PLC manufacturers use switch assemblies in the I/O rack to number the racks. A typical programmable controller I/O rack selection switch arrangement is shown in Table 11-1.

## Control Panel Specification

In most cases, the equipment enclosure for a programmable controller system is fabricated by an outside control panel vendor, so that an equipment specification will be required to cover all requirements of a system. The following is a typical control panel specification:

The control panel furnished under this specification shall be supplied complete with the instruments and equipment listed on the enclosed drawings, installed and electrically wired, and ready for wiring to field instruments and equipment.

The control panel shall conform to the following:

1. The control panel shall be fabricated with cold-rolled steel plate.
2. All miscellaneous items, such as wire raceways, terminal strips, electrical wire, etc., where possible, shall be made of fire-resistant materials.

| I/O Rack Number | Switch Position | | | |
|---|---|---|---|---|
| | 4 | 3 | 2 | 1 |
| 00 | Closed | Closed | Closed | Closed |
| 01 | Closed | Closed | Closed | Open |
| 02 | Closed | Closed | Open | Closed |
| 03 | Closed | Closed | Open | Open |
| 04 | Closed | Open | Open | Open |
| 05 | Closed | Open | Open | Closed |
| 06 | Closed | Open | Open | Open |

Table 11-1. Typical rack number selection table.

3. The grounding bus bar and studs shall be pure copper metal.

4. A minimum of two 120-V ac, 60-Hz utility outlets shall be installed in panel.

5. An internal fluorescent light with a conveniently located ON/OFF switch shall be installed in the panel.

6. A circuit breaker panel with circuit breakers to accommodate all panel lighting, power supplies, instruments, I/O modules, processors, and any other loads listed on the system drawings shall be provided.

7. Each electric wire over 12 inches in length shall be identified at each end with a wire number per the electrical drawings for the system.

8. Terminal strips with screw-type connectors shall be used and no more than two wires shall be terminated on any single terminal.

9. Nameplates shall be made of laminated plastic with white letters engraved on a black background for both front and rear panel-mounted instruments and components, such as power supplies, transformers, and PLC I/O racks.

10. The ac wiring shall be separated from 4 to 20-mA dc current and digital signals by a minimum of 24 inches, and they must be wired to separate terminal strips.

11. All electrical wires and cables shall enter the control panel through the top. Sufficient space shall be provided to allow ac power cables entering the top of the panel to continue directly to the circuit breaker panel.

12. All wires and cables shall be routed and tied to provide clear access to all instruments and components for maintenance and removal of defective components.

13. All dc wires shall have a minimum insulation voltage rating of 600 volts ac, and all dc wires shall have a minimum insulation voltage rating of 300 volts dc.

14. All electrical conductors shall be copper of the correct wire size for the current carried with 98% conductivity, referenced to pure copper.

15. Wireways shall be attached securely to the control panel.

16. Metal surfaces with wires or cables passing through them shall be furnished with insulated polyethylene grommets to prevent damage to the conductors or cables.

17. All wire bundles or cables shall be clamped to the panel at all right angle turns.

18. All wires entering or leaving a wire bundle shall be tied to the bundle at the point of entering or leaving the main wire bundle.

## Equipment Layout Design

Proper control system equipment layout design can reduce installation costs and improve system reliability and maintainability.

In addition to the programmable controller components, the equipment layout must also take into account the other system components, such as field devices and instruments, power disconnect boxes, power transformer, and the location of process equipment and machines.

In general, placing the processor near the process equipment and using remote I/O racks where possible will reduce wire and electrical conduit runs. It is possible that the cost of wiring and conduit installation will far exceed the programmable controller equipment costs if care is not taken in the equipment layout design.

The control panel should be placed in a position that allows the doors to be opened fully for easy access to wiring terminals and system components for maintenance and troubleshooting. The National Electric Code (NEC) requires that the panel doors open at least to 90°, and there must be a minimum of 36 inches of clearance from the rear of the panel to the nearest grounding surface or wall. Before starting the design of a control panel, the designers should read and understand section 110 of the National Electrical Code to avoid any code violations or safety problems.

An emergency disconnect switch should be mounted on or near the control panel in an easily accessible location.

If the location where the control panel will be placed contains equipment that generates excessive radio frequency interference (rfi) or electromagnetic interference (emi), the panel should be placed a reasonable distance from these sources. Examples of such sources include electric machinery, welding equipment, induction heating units, and electric motor starters.

# System Start-Up and Testing

The first requirement for successful system start-up is to have a written system operational (SO) test procedure. This procedure should be written by the control system designer and carefully reviewed and approved by all parties involved in the project. A time-saving procedure is to functionally test the control system program using a process simulator before the system is installed on the process.

The SO test procedures will generally contain the following main sections: visual inspection, continuity test, input signal testing, testing of outputs, and process operational testing. A typical SO test procedure follows.

## Typical SO Test Procedure

### I. Visual Inspection

1. Verify that all system components are installed per the system drawing.
2. Check I/O module location in equipment racks per I/O drawings.

3. Inspect switch settings on all intelligent modules per drawings.
4. Verify that all system communication cables are correctly installed.
5. Check that all input wires are correctly marked with wire numbers and terminated at correct points on input module.
6. Check that all output wires are correctly marked with wire numbers and terminated at correct terminals on output module.
7. Verify the power wiring is installed per the ac distribution drawing.

## II. Continuity Check

1. Use ohmmeter to verify that no ac wire is shorted to ground.
2. Verify continuity of ac hot wiring.
3. Check ac neutral wiring system.
4. Check continuity of system grounds.

## III. Input Wiring Check

1. Place the programmable controller in program mode.
2. Disable all output signals.
3. Turn "on" ac system power and power to input modules.
4. Verify the E-stop switch removes power from the system.
5. Activate each input device, observe the corresponding address on the programming terminal, and verify that the indicator light on the input module is energized.

## IV. Output Wiring Check

1. Disconnect all output devices that might create a safety problem, such as motors, heaters, or control valves, etc.
2. Place programmable controller in test mode.
3. Apply power to the programmable controller and the output modules.
4. Depress the E-stop push button and verify that all output signals are deenergized.
5. Restart system and use the forcing function in the programming terminal to energize each output individually. Measure the signal at the output devices and verify that the output light on the module is energized for each output tested.

### V. *Operation Test*

1. Place processor in "Program" mode and turn on main power switch.
2. Load the pre-tested control program into the programmable controller.
3. Disable all outputs, select the run mode on the processor, and verify that the run light on the processor is activated.
4. Check each rung of logic for proper operation by simulating the inputs and verifying on the programming terminal that the correct output is energized at the proper time or sequence in the program.
5. Make any required changes to the control program.
6. Enable output modules and place processor in "Run" mode.
7. Test control system per process operating procedure.

# Maintenance Practices

Programmable controller components are designed to be very reliable, but occasional repair is still required. System maintenance costs can be greatly reduced by using good design practices, complete documentation, and preventive maintenance programs.

## Preventive Maintenance

A systematic preventive maintenance program will reduce the down time of the control system. The preventive maintenance of programmable controller components is usually scheduled at the same time as the machine or process that is down for maintenance or repair. Normally, since programmable controller equipment is more reliable than some machinery or process equipment, it requires less frequent preventive maintenance operations.

It is very important in a preventive maintenance program to carefully check the various connectors in the system. It is estimated that 70% of the problems found in electronic or computer-based systems are caused by loose, dirty, or defective connectors. The connection to the I/O modules should be checked periodically to make sure no wiring has come loose. The seating of the I/O modules in the equipment rack should also be checked. If the system is located in process areas that have high vibration levels, the preventive maintenance check should be performed more often.

Excessive heat is another major cause of failure in programmable controller systems. Therefore, if a enclosure is cooled with fans, the filters used on the panel must be cleaned or replaced on a regular basis.

It is important that dirt and dust is not allowed to build up on system components, because a dirt buildup on electronic components can reduce heat dissipation and cause an overheated condition in the system.

Electrical noise can cause erratic and dangerous operation of a PLC system, so the maintenance department must check to make sure that equipment producing high levels of rfi or emi noise are not moved near the programmable controller equipment.

Maintenance personnel should also check to make sure that unnecessary items are not stored on or near the PLC equipment. Leaving items such as tools, test equipment, drawings, or instruction manuals in the control panel can obstruct the air flow and cause heating problems.

An important maintenance practice is to stock spare parts for the system. This reduces down time in the event of a component failure.

## Troubleshooting

Troubleshooting can be defined as the methods used to determine why a system or component is not functioning properly. Troubleshooting, as with many practical skills, is an art as well as having an analytical or logical basis. As such, troubleshooting procedures are a trainable skill. Figure 11-3 shows a flow diagram for the basic analytical methodology that should be used to solve any problem.

The first step in troubleshooting is to identify the problem. This is typically a description of the problem by the person reporting the problem and consists of symptoms of the problem and the reporting party's views on the problem.

The next step is to collect information about the problem. This data gathering includes questioning of the problem reporter for more detailed information, viewing physical symptoms, and reviewing the components and systems involved.

The third step is to narrow down the problem. If the problem involves complex, interactive components, try to narrow it down to the component that is causing the problem. Once the problem has been narrowed down, apply a troubleshooting method or methods to solve the problem. If successful, correct the component; if not, return to the appropriate step and start again.

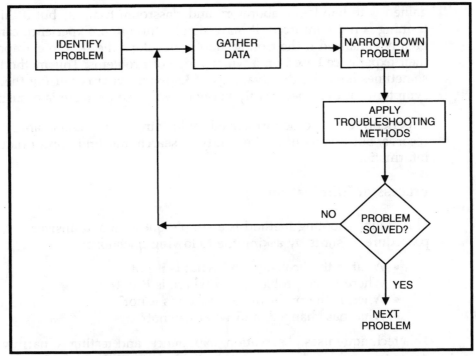

**Figure 11-3. Troubleshooting flowchart.**

## Troubleshooting Method

There are many different methods of troubleshooting and each has its advantages and disadvantages. The approach used is often chosen due to a personal preference or convenience, but in some cases it is dictated by the problem itself. In most cases, more than one method will be used and sometimes even intermixed.

We will discuss the following six main troubleshooting methods: (1) experience, (2) process of elimination, (3) divide and conquer, (4) remove and conquer, (5) substitution, and (6) setting a trap.

### Experience

This is the most common method and many times the simplest method. You know the problem and its solution because you have seen or heard of it before. If there are a number of possible solutions, experience can be used to find the initial area to attack first. The selection can be based on the highest probability, the one with the least risk or upset, the easiest, the closest, the most easily testable, etc.

Experience is primarily an on-the-job learned skill: the more work you've done, the more experience you have. Experience can also be gained with hands-on laboratory and classroom training, but if this training is not reinforced, it may soon be forgotten. Experience can be greatly enhanced if the troubleshooter develops the skill to extrapolate their experience based on a wider range of problems. This method can sometimes have the disadvantage of knowing what is causing the symptoms but not necessarily knowing why the problem is occurring.

Experience can also be formalized by keeping good maintenance records, but this requires the ability to search and find appropriate information.

## Process of Elimination

This troubleshooting method is a simple question and answer procedure. It starts by asking the following questions:

- What is the problem, and what is it not?
- Where is the problem, and where is it not?
- When is the problem, and when is it not?
- What has changed, and what has not?

This technique uses observation, experience, and testing to narrow down the problem into a more workable form. This method is not always linear; that is, the next troubleshooting step may not depend on the prior step.

## Divide and Conquer

This method is commonly taught in technical schools and courses. It consists of dividing or breaking the system down into two parts, testing, and finding out which is working properly and which is not. The part that is not working is further divided and testing is applied again until the cause of the problem is obtained, as shown in Figure 11-4.

The secret to this method is knowing where to divide the system. The dividing point may be based on experience, ease of access, test points, etc. When in doubt, divide it in half. This method works well in PLC systems because you can easily divide the system components into field devices, I/O modules, CPU, or control software.

## Remove and Conquer

This method works best with loosely coupled systems. An example might be several programmable logic controllers (PLCs) connected on a communication link. If the data are being corrupted, removing each PLC in turn may help determine if one of the PLCs is to blame.

In complex systems, many discrete modules and electronic circuit cards are called on to do different and/or similar functions. A method to

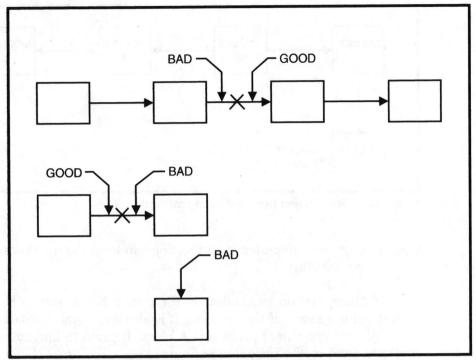

**Figure 11-4. Divide and conquer troubleshooting method.**

solving problems in these types of systems is to remove the modules or cards one at a time from the system and see if the problem goes away. Alternatively, remove all the modules or cards and add them back in until the problem comes back. This method is very effective in modular PLC I/O systems. A block diagram to illustrate this method is given in Figure 11-5.

## Substitution

This method consists of substituting a known good part into the system to see if the problem goes away. This is typically used on complex,

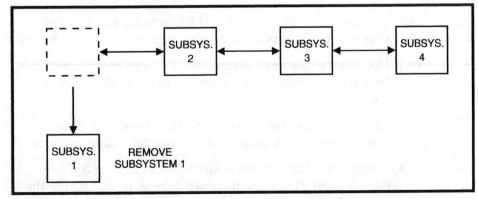

**Figure 11-5. Remove and conquer troubleshooting method.**

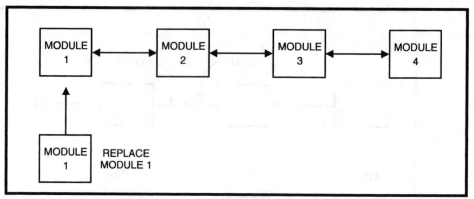

**Figure 11-6. Substitution troubleshooting method.**

black-box systems. Experience and testing can limit the number of boxes to be substituted.

A good conceptual understanding of the system being troubleshot is critical to the success of this method. It is also the method of last resort for subtle or intermittent problems. A block diagram to illustrate this troublshooting method is given in Figure 11-6.

### Setting A Trap

Setting a trap is often the only effective method for catching the cause of a spurious or transient problem. With the advanced PLC systems that include data logging, archiving, and trending, much more data is available than in the past. Even so, this sometimes is not enough and additional monitoring points or traps are needed to catch the "prey" in question. This method is very effective in software troubleshooting where the program can be halted if a certain condition or conditions are detected.

### EXERCISES

11.1  Discuss the main reasons and advantages for placing programmable controller components in metal enclosures.

11.2  Discuss the need for designing control panels to meet the heat requirements of programmable controller system components.

11.3  List some important design practices used to reduce maintenance costs on programmable controller systems.

11.4  Design and draw a power distribution system for the dehydration PLC system shown Figure 10-8. Assume the system is mounted in a control panel and has a 6 socket ac

outlet power strip, two cooling fans, and a single
fluorescent light.

11.5 Discuss the importance of proper grounding techniques in
programmable controller design.

11.6 List the various phases of a typical system operational test
procedure.

11.7 Discuss the important features of an effective preventive
maintenance program for programmable controller
systems.

11.8 Discuss the common troubleshooting methods used in
finding problems with PLC systems.

## BIBLIOGRAPHY

1. *Assembly and Installation Manual PLC5 Family Programmable Controllers*, Allen-Bradley, Publication 1785-6.6.1, November, 1987.

2. Bryan, L. A., and Bryan E. A., *Programmable Controllers: Theory and
   implementation*, Industrial Text Co., 1988.

3. *Assembly and Installation Manual PLC-2/20, PLC-2/30 Programmable
   Controllers*, Allen-Bradley, Publication 1772-6.6.2, March, 1984.

4. *National Electrical Manufacturers Association (NEMA)*, Enclosures for
   Electrical Equipment (1000 Volts Maximum), NEMA Standard Publication No. 250, 1979.

# INDEX